Navy Daze

NAVY DAZE

Coming of Age in the 1960s
Aboard a Navy Destroyer

Michael R. Halldorson

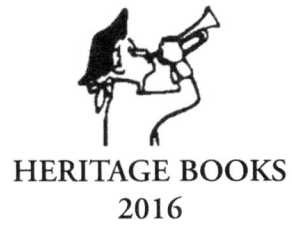

HERITAGE BOOKS
2016

HERITAGE BOOKS
AN IMPRINT OF HERITAGE BOOKS, INC.

Books, CDs, and more—Worldwide

For our listing of thousands of titles see our website
at
www.HeritageBooks.com

Published 2016 by
HERITAGE BOOKS, INC.
Publishing Division
5810 Ruatan Street
Berwyn Heights, Md. 20740

Copyright © 2016 Michael R. Halldorson

All rights reserved. No part of this book may be reproduced or transmitted in any form or by any means, electronic or mechanical, including photocopying, recording or by any information storage and retrieval system without written permission from the author, except for the inclusion of brief quotations in a review.

International Standard Book Numbers
Paperbound: 978-0-7884-5735-7
Clothbound: 978-0-7884-5951-1

To my father, TSgt Magnus Michael Halldorson, United States Army; my brother Alan, a Navy veteran who served on tank landing ships in Vietnam; and to my nephew Brian Halldorson, who more recently served a tour in the "modern" Navy. My father's duty was the most distinguished. He was twice wounded during the Battle of the Bulge and captured; and later, while in a German Prisoner of War camp, Stalag IX-B, almost died of malnutrition. My brother's service, like my father's, was also memorable. Alan served aboard the tank landing ships USS *Caddo Parish* (LST-515) and USS *Iredell County* (LST-839) in the dangerous, confined inland waters of the Republic of Vietnam. Both these ships took rocket-propelled grenade and automatic weapons fire from the Viet Cong. Much thanks is due my dear wife Merry for her love and support, my sister Jan MacDonell for her encouragement to write this account, and my mother, Louise Halldorson, who was always my greatest advocate.

Contents

Foreword	ix
Acknowledgements	xi
Preface	xiii
1. Naval Reserve Duty	1
2. Boot Camp	3
3. Officer Candidate School	7
4. Active Duty	13
5. Life Aboard the *Hopewell*	21
6. Initial Portion of the Deployment	37
7. Latter Part of Deployment/Combat Duty	45
8. A Long-Awaited Visit Home	71
9. Liberty in San Diego and Long Beach	77
10. Second Deployment	85
11. Ship's Office/Captain's Mast	111
12. Return to San Diego/Overhaul	117
13. Return to College Life and Duty in the Naval Reserve	121
14. *Hopewell* Reunions	127
Postscript	133
Appendix: My Time with Escher	139
About the Author	145

Photos and Images

4-1: My '56 Chevy was not a hotrod	16
5-1: *Hopewell* under way	22
5-2: The author in 1964	23
5-3: Girls back home	35
7-1: *Hopewell* firing her 5-inch guns	51
7-2: Photograph of an envelope displaying artwork	54
7-3: Dungaree jacket patches	55
7-4: Spent brass shell casings returned to their canisters	57
7-5: A postcard hand drawn for Christmas in 1964	59
7-6: *Hopewell* and *Bon Homme Richard* refueling from *Mattaponi*	66
7-7: Cartoons from the *Hopewell*'s cruise book	67
8-1: Barracks at Stalag IX-B, Bad Orb, Germany	74
9-1: Two-tone 1955 Buick	79
9-2: "Whiskey a' Go-Go" Night club in Los Angeles	80

9-3: The Seven Seas locker club in downtown San Diego 81
10-1: Two sailors at Enlisted Man's Club in Subic Bay 87
10-2: 5"/38 gun mounts 89
10-3: Victim of an unprovoked assault in a bar 93
10-4: Viet Cong weapons captured by U.S. Marines 95
10-5: Vietnamese Coastal Force craft 96
10-6: The "Bear Hole" in upper Bidwell Park, Chico, California 102
10-7: F-4 Phantom launching from the carrier *Constellation* 104
10-8: A-6A Intruder aboard the carrier *Forrestal* 105
10-9: Destroyer *Vance* under way off Oahu, Hawaii 109
11-1: Yeoman at work in *Hopewell*'s ship's office 113
13-1: Party scene in the late 1960s 123
13-2: Linocut print of *Hopewell* 124
14-1: Photograph of the bow of the ex-*Hopewell* 131
Appendix-1: Escher at work 140
Appendix-2: Halldorson print *At That Point in Time* 143
Appendix-3: Halldorson print *Babylonian Pipe Dream* 144

Maps and Diagrams

6-1: Philippine Islands 41
7-1: Indochina 49
7-2: 5-inch gun round projectile and powder 62
7-3: Gun barrel and breech of a 5-inch/38-caliber mount 63
7-4: Indicators for training 5-inch gun mount 64

Foreword

Venturesome men, and some women, have for centuries gone down to the sea in ships, seeking adventure and the chance to travel to exotic foreign lands. For some, life aboard a ship at sea provided an escape from deprivation, or a new start in life. Today, young men and women still join the Navy, Coast Guard, or Merchant Marine seeking adventure, or to better themselves. Many find life at sea hard, and sometimes lonely, and choose to leave the sea service following their "hitch." Life passes quickly, and most, as the years pass, find themselves reminiscing about sailing deep waters, runs ashore, and good times spent with shipmates.

Every former sailor—whether they are from the "greatest generation" of World War II, or of more recent vintage—will find many commonalities between their experiences and those set down by Mike Halldorson in *Navy Daze*. It's difficult for me to harken back to my enlisted time before becoming an officer, because much of my former carefree life vanished upon my receiving a commission, but Mike's book takes me right back to reporting aboard my first ship as a seaman recruit. The life of any seaman aboard ship is pretty much the same; eight hours of watch a day, with additional time devoted to such things as refueling, flight operations, and gunnery practice. Moreover, the work day is not necessarily over when one gets off watch. Seamen get the tasks that senior enlisted do not want to do, and are exempt from, by virtue of working their way up from seaman.

Every Navy ship counts amongst its crew at least one bully, at least one individual willing to do almost anything to get some laughs, a few "scrounges" that need encouragement to take showers, and a few slackers willing to let others pull their load. Scrounges and slackers are usually shown the error of their ways by shipmates, sometimes not very gently. Most of the men, and today women as well, who comprise a ship's crew are the salt of the earth. They work hard, take pride in their work, take care of their shipmates, and have a good time on the beach when in port. Sailors that particularly like to "bend an elbow" are known as "steamers." Despite the popular image that all navy men are drunken sailors, many like to engage in athletics during their free time, or pursue tourist opportunities when overseas. Most enjoy a combination of these activities, and have their wits about them when it is necessary to help a shipmate who has overindulged.

Despite life at sea often being hard and spare, some sailors choose to remain in the Navy at the end of their enlistment by reenlisting, which is known as "shipping over." The following poem (by an unknown author) aptly expresses the wishes of one individual to do so. The term "mast" refers to "Captain's Mast," a procedure whereby the commanding officer makes inquiry into the facts surrounding an offense allegedly committed by a member of the command; affords the accused a hearing as to such offense; and disposes of such cases by dismissing the charge, imposing punishment, or referring the case to higher authority for a court-martial. Standing at attention in front of one's commanding officer is not something that most sailors enjoy. "Request" refers to a "request chit." Sailors wishing to remain in the Navy must request this action, and their command must approve it in order for them to remain in the naval service.

>Our Father who art in Washington
>Please, dear Father, let me stay,
>Do not drive me now away.
>Wipe away my scalding tears
>And let me stay my thirty years.
>Please forgive me all my past
>And things that happened at the mast,
>Do not my request refuse,
>And let me stay another cruise.

I, like Michael Halldorson, chose to leave the Navy at the completion of my enlistment; in my case to complete a college degree, and hopefully earn an officer's commission that would someday lead to command of a ship at sea. Mike returned to Chico, a hometown we share, studied art in college as both an undergraduate and graduate student, and pursued business and art careers. Today, he is a board member of the Janet Turner Art Museum, a member of Rotary and other local civic groups, and the commander of VFW Post 1555. The Navy served him and me well.

David D. Bruhn
Cdr. USN (Ret.)
Former commanding officer, USS *Gladiator* (MCM-11) and USS *Dextrous* (MCM-13)

Acknowledgments

First of all, a tip of the hat to Senior Chief Data Processor Daniel Smith, USNR (Ret.) for providing me a photograph of a pair of beautiful silk "liberty cuffs" depicting dragons for use as the backdrop for the cover art. When visiting the Orient, many sailors had this kind of decorative material sown inside the sleeve cuffs of their uniform jumpers. Although officially forbidden, most officers would overlook sailors folding back their cuffs to display the silk linings as long as such action occurred only inside a club or bar, and not on the street in public. The cuffs had great value to their owners back home in the United States, as such adornment conveyed that their owners were "old salts"; particularly in comparison to peers who did not have readily visual proof of having been overseas. Smith's collection of photographs of various liberty cuffs may be viewed at his website: http://navydp.com/NavyCollector/Navy_Traditions.htm.

    I am indebted to my former shipmates and now lifelong friends Jim Brickey and Johnny Junior Sharp. As related to this book, they refreshed my memory on our shared experiences aboard ship and during runs ashore, and provided me photographs for use herein. Another shipmate, R. Mike Sohikian, shared valuable information on various operations of the *Hopewell* (DD-681).

    I am also grateful to Lt. George Cooper and Lt. (jg) Gustav R. Scherer, *Hopewell*'s operations officer and navigator, respectively, during my tour aboard her. Lt. Comdr. James H. Burpo, one of the three executive officers assigned to the ship during my tour of duty, assisted me greatly by sharing his perspective as well. The official ship's history was, of course, very important in providing context by linking the anecdotes and episodes I have recounted.

    I am much appreciative of the assistance lent me by Comdr. David D. Bruhn, USN (Ret.). Bruhn is a prolific author of books on naval history and is currently working on his ninth book. He generously reviewed drafts of the manuscript and made suggestions for changes.

    Finally, Lynn Marie Tosello—a consummate professional and editor extraordinaire—contributed much improved diction, as well as eloquence and humor to the book. She was also instrumental in advising me when certain nautical terms, or "fleet short-hands" familiar to former sailors, needed explanation for those who "go down to the sea in ships" vicariously.

Preface

Twenty years from now you will be more disappointed by the things that you didn't do than by the ones you did do. So throw off the bowlines. Sail away from the safe harbor. Catch the trade winds in your sails. Explore. Dream. Discover.

—Mark Twain

Navy Daze is a story of a young man from a small northern California town who, with no clear direction in life, joined the Navy and came of age aboard a destroyer during the Vietnam War. In fact the book might be described as *American Graffiti* meets *Apocalypse Now*. For those not familiar with these two acclaimed movies, the first is about the wholesome lives of American high school students and the latter, the brutal Vietnam experience of those who served in the U.S. military action to preserve freedom in the Republic of (South) Vietnam. Former navy men, of any generation, will likely find they, too, have had many of the experiences described herein. I did not do anything heroic during my tour aboard the destroyer USS *Hopewell* (DD-681). The *Hopewell* and her crew did carry out all assignments, including those on the gunline off South Vietnam.

For those readers not familiar with the unique characteristics and culture of America's military services, I will offer a short explanation. When the United States is not at war, its ground and air forces—the U.S. Army and the Air Force—are essentially garrison forces that normally remain on their bases except for periodic training exercises. The Navy and Marine Corps function as deterrent forces—"policemen on the beat," so to speak. Aircraft carriers and their escort ships, and groups of amphibious ships loaded with Marines, patrol the world's oceans to help maintain the peace for America and its allies. Young men and women who join the Navy and serve aboard ships, usually have opportunity to travel overseas and experience cultures and environments very different from those to which they are accustomed. I did so and, because my country was at war, I experienced combat duty as well.

Before I joined the Navy, I, like many of my peers, was interested in girls, cars, and alcohol. After my stint in the Navy, I was still interested in girls, cars, and alcohol, but my horizons had widened thanks to the maturity I gained while in the service. After my Navy hitch, I reentered college, did well in my studies, and pursued other interests as well, including what some might consider a dichotomy, creating fine art while a member of the Naval Reserve.

My experiences were common to many "bluejackets" (enlisted sailors) from that era. Two months after I reported aboard the *Hopewell*, she stood out of San Diego en route to the Republic of Vietnam. Time spent on the gunline in the combat zone off South Vietnam was interspersed with periodic liberty for the crew in Japan, Hong Kong, and the Philippines. *Hopewell* returned to San Diego following this deployment, and later made another one to Vietnam while I was aboard her.

Vietnam changed me in many respects. In January 1968, the North Vietnamese and Viet Cong launched the bloody Tet Offensive, which spurred much dissent, on American college campuses and in the streets, regarding continued United States involvement in the war. Though I had completed my enlistment and returned to college studies before this event, I experienced the same hostility as many veterans returning home; many countrymen were passionately opposed to the war. The general attitude of the nation has changed since that time; today most people believe that, while you may not support your country's involvement in a war or conflict, you should support the brave men and women who fight for freedom.

People tend to remember the good times and forget particularly difficult or unsettling experiences. Sailors are no different. When I visit with or call one of my old shipmates, the conversation almost always revolves around fond memories with an occasional "what about?" thrown in. If I chose to, I could bring to the fore vivid memories of having my nose broken in Japan and a tooth sheared off in the Philippines; both incidents took place in drinking establishments ashore. Although I have included accounts of these incidents, I prefer to remember enjoying the company of a lady in Hong Kong or tours of the Tiger Balm Gardens in that city, Manila's Royal Palace, and Japan's Mount Fuji. Pushed to the back of my mind are such things as trying to survive particularly rough seas, and hours-on-end spent inside a hot and humid five-inch gun mount while patrolling off the coast of South Vietnam.

Whether viewing the expansive Pacific, or experiencing the violent shaking and stench from the firing of my ship's guns, I felt an incredible sense of pride at being in the Navy. It is my hope that readers enjoy this book, and that it helps to bring back fond memories of youth, distant waters, and good shipmates for those who trod the deck of a ship.

The mistakes I made as a young man helped to shape the older and wiser individual that I am today. Many if not most young men are immature, and young sailors overseas had ample opportunity to get into trouble, while pursing adventure. I had my share of gaffes, as readers will deduce, but feel well-served by my Navy experiences.

NAUTICAL TERMS

Some readers may find useful the definitions of some nautical terms sprinkled through the book:

- Atoll: A ring-shaped coral reef or a string of closely spaced small coral islands, enclosing or nearly enclosing a shallow lagoon. The largest island of an atoll often has the same name as the atoll, just as the largest island of an island chain often has the same name as the chain.
- Broach: To turn the ship broadside to heavy seas, or lose control of steering in following seas so that the ship is turned broadside to the waves. An extremely dangerous situation in steep seas since the ship may roll over and capsize.
- Caliber: The bore-to-barrel-length ratio of a naval gun, obtained by dividing the length of the barrel (from breech to muzzle) by the barrel diameter to give a dimensionless quantity. For example, a 3-inch/50-caliber gun has a barrel length of 150 inches.
- Conflagration: Large and destructive fire.
- Deckhouse: An enclosed structure built on the ship's upper or main deck, usually the navigating station though the term can refer to any simple superstructure on deck.
- General Quarters: Battle Stations.
- Rating: The rating of a sailor is a combination of rate (pay grade, as indicated by the number of chevrons he or she wears) and rating (occupational specialty, as indicated by the symbol just above the chevrons).
- Slew or slewing: Moving a gun mount side-to-side (train) or up and down (elevation).
- Stand (past tense stood): Of a ship or its captain, to steer, sail, or steam, usually used in conjunction with a specified direction or destination, e.g., "stand into port."

- "Tin can": A common nickname for a destroyer. The nickname arose because in World Wars I and II, the hull plating of this ship type was so thin the sailors claimed they were made from tin cans. In fact, a .45 pistol bullet would penetrate it. Modern destroyers have much thicker hull plating, but the nickname persists.
- Vessel: Any craft (from largest ship to smallest boat) that is capable of floating and moving on the water.

Famous World War II recruiting poster. During my time in the Navy, both men and women could serve—a practice which continues today.

1

Naval Reserve Duty

I can imagine a no more rewarding career, and any man who may be asked in this century what he did to make his life worthwhile, I think can respond with a good deal of pride and satisfaction: "I served in the United States Navy."

—President John F. Kennedy

Following my high school graduation, a Navy recruiter, Chief Machinist Mate John Hammons, contacted me in September 1961. Hammons' bearing in his dress blue uniform was very impressive. He of course explained the benefits of being in the Navy versus the other services, and posed a question: If nuclear war broke out, would you rather be on board a ship in the vast expanse of the ocean, or in a foxhole? Having come of age during the *Cold War* with the Soviet Union, with the attendant nuclear attack drills in which school children took shelter under their desks, a ship sounded better to me. "Where do I sign?" I responded, and so began my naval career.

I must explain that as a high school student in the late 50s and early 60s, I was completely naïve about world affairs and did not realize that actions were taking place in Southeast Asia that would directly affect my life a few years later. United States military advisors had already been sent to Vietnam and the "Domino Theory" was shaping national policy. President Dwight D. Eisenhower had coined this phrase. He suggested that, should French Indochina (Vietnam, Laos, and Cambodia) fall to the communists, it could create a "domino" effect in Southeast Asia. The so-called "Domino Theory" would dominate U.S. thinking about Vietnam throughout the 1960s.

I enlisted in the Naval Reserve, committed to being the very best sailor who ever lived. Move over John Paul Jones. One of the first formal activities was a personnel inspection conducted at our Reserve unit at the Chico, California airport. We were in our dress blues with our shoes "spit shined." I was a seaman recruit (E-1), the lowest of low; still I had one white stripe on my sleeve. (This convention no longer exists in the Navy; nowadays, seaman recruits display no rank.) Seasoned reservists told me that I was so low that "whale sh*t would look like stars!" to me. That did not diminish my enthusiasm. I was ready to show the admiral conducting the inspection what a squared away seaman recruit looked like. Instead of a depiction of "Uncle Sam" in the famous recruiting poster pointing at prospective enlistees, it would be me!

I was in the second row of the formation. It was really warm and as I stood at attention, I could feel a little sweat running down my arms and around my neck. My last thought was, "What is taking that admiral so long to get to me?" He only had two ranks of sailors to review before my group. When I came to, I was lying on the floor with my fellow reservists huddled around my prone body. In an effort to present a picture-perfect pose, I had locked my knees, passed out, and fallen on the guy directly in front of me. I had really wanted to impress the admiral, and likely had, but not in the way I had intended. Years later when my younger brother Alan was at an inspection at our Naval Reserve Unit, he fell on the guy in front of him; just as I had done years before. He also locked his knees, which seemed to be a Halldorson trait. Years later when his son, my nephew Brian, joined the Navy, he did not suffer such indignation.

I was a naval reservist for ten months before I was sent to San Diego for Naval Recruit Training ("boot camp") in July 1962.

2

Boot Camp

The weeks of training in drill and ceremony are put to the task during Graduation/Pass and Review. Your dress uniform, specially tailored for you is donned, your shoulders are square and your head high. You have accomplished a task few in the US will and are now part of the world's finest naval force.

—Military Readiness Command guidance

Boot camp in San Diego was hell—all two weeks of it. (At least that was my youthful perspective at the time.) Had I been a "regular Navy type" it would have been eight weeks. As depicted in films and literature, recruit training was characterized by much time spent spit shining shoes, marching, attending courses of instruction, studying, early morning wake up calls (called reveille), eating chow with hundreds of your new best friends, and getting demerits for something you did, or did not do. As an American teenager, I was used to getting up late – boot camp required rolling out of your bunk at 4 a.m.

Marching drills and book training were the order of the day, interspersed with things like firing weapons and fighting black oil fires. We carried fire hoses into a building made of cinder block and put out fires meant to simulate an engine room conflagration aboard a ship. The only protection from smoke, for me and several other guys who were using a high-pressure fire hose to extinguish the flames, were our knit watch caps. Then there was tear gas training. We were given gas masks, herded into a room that was then filled with tear gas, instructed to remove our masks, and recite the "Gettysburg Address!" If any missed one word, he would have to start over. Well, that is not exactly factual; I'm pretty sure that it was name, rank and serial number, but the time spent in the chamber seemed like an eternity. The training with firearms—small-caliber rifles—involved instruction in their use and target practice.

I had one particularly surreal experience during Boot Camp. For some reason that eludes me now, I was summoned to the administrative office. As I was walking by the WAVES barracks the windows were fully open on that very hot August day due to the lack of air-conditioning. WAVES was a World War II acronym for "Women Accepted for Volunteer Emergency Service," and was then still commonly used in reference to women in the Navy. I heard a woman with a commanding voice yell, "Attention," followed by a directive that I will not repeat here. To say the least, I was shocked that a woman would speak like that to other women, or for that matter, that any woman would express herself thusly. Perhaps she was trying to be "saltier" than her male counterparts, or was particularly foul mouthed. To this day, such language is not what I would expect from a female recruit company commander "pushing boots" (recruits). This puritan perspective that I hold today, may in part be due to my age.

It is customary that recruits be granted "boot camp liberty" the day before graduation to reacquaint them with the civilian world before completing training and reuniting with family and friends. While enjoying this liberty, a buddy and I went to a burlesque show in downtown San Diego in which ladies stripped on stage. Later, we went to a movie theater where an unclothed woman on screen writhed around on a bed for thirty minutes! Ahhh, for a young sailor away from home for the first time, that was livin'.

However escaping boot camp was not as easy as returning from liberty and going through a ceremony. The morning of our last day, we had to stand a final inspection administered by the company commander. He strode through our ranks (lines of sailors at attention) and could not find anything wrong. Everyone's uniform was neat and their shoes spit-shined, but therein was the problem; he had to find something askew. An example had to be made of someone. As he came to the end of the last row, who do you think was standing there at perfect attention? Yes indeed it was me, and this time I had not locked my knees. He looked back at the formation, cocked his head, ran his eyes up and down the side of my face, and said, "This one needs to get closer to the razor, forty jumping jacks!" So in front of my company, I and the guy next to me who had not applied black polish to the welts of the soles of his shoes, or so he was told, did calisthenics.

On my trip back to Chico, I did not quite enjoy the homecoming I had expected from my girlfriend. I returned home with a new swagger and my regulation military haircut, which had grown out to about one-quarter inch. Where my white hat had shielded me from the sun over the past two weeks, my forehead was white; and where the sun hit many hours a day, I had lost several layers of skin. I thought my appearance and demeanor pretty cool, but my sweetheart could hardly stand to look at me. To be fair to her, I was not returning from the war, I had only been gone two weeks, and I was less handsome as a result of my short stint with the Navy.

I settled back into Chico life and continued to be a naval reservist ("weekend warrior") while attending Chico State College. A year later, I would attend Officer Candidate School in Newport, Rhode Island—the prospective officer's version of Boot Camp.

3

Officer Candidate School

Those who expect to reap the blessings of freedom must undergo the fatigue of supporting it.
—Thomas Paine

The summer of 1963 found me attending Reserve Officer Candidate School (ROCS) in Newport, Rhode Island, as a result of a particularly strong recommendation from Capt. Pete Volpato, USNR, the commanding officer of my naval reserve unit in Chico. Following arrival at the training command, I was placed in Company Romeo Three. As one might expect, it was made up of guys from all across the country. Some had interesting idiosyncrasies, including a fellow who would sing profane little ditties after the lights went out. He had one particularly raucous song that he sang to the tune of "These Foolish Things Remind Me of You." I cannot repeat his lyrics in this book. The barracks howled and he had his moment each night. He talked with a thick Bronx accent, so we assumed he was a New Yorker. He was actually from New Jersey; both sound the same to someone from California. To this day, that raunchy little number gets stuck in my head as I am doing mundane things like yard work.

We also had a cadet who had attended the Citadel, the Military College of South Carolina, located in Charleston. He knew how to count cadence, military style, and we marveled at his military bearing. There was also a "surfer dude" from southern California with a deep, deep orange-brown tan. He used a lot of slang unfamiliar to me, and frequently uttered the word "Dude," which is currently in vogue, but was not back then. Nothing seemed to be very important to him; he was very "laid back." The Safaris' song "Surfer Joe" from 1963, fit him to a tee, except that he was short.

Down in Doheny where the surfers all go,
There's a big, bleached blondie named Surfer Joe.

> He has a green surfboard with a woody to match,
> And when he's ridin' the freeways—
> Man, is he hard to catch!

Finally, among the cadets most memorable to me was Chris Dulis from Portland, Oregon, who told me to keep my ears open for a great new song that had not hit the air waves yet, titled "Louie, Louie." He would become a good friend. I should perhaps explain that I, like many of my peers, was crazy about rock & roll music—as well as girls, hot rods, and alcohol.

Our time at ROCS was largely devoted to classroom instruction, marching and drill. We did not do a lot of "PT" (physical training) because of the heavy academic load intended to prepare us to be naval officers. Seamanship was taught by a regular Navy lieutenant who did not try to mask his disdain for reservists. He had a swagger, a Navy regulation haircut, and appeared to be "squared away." He probably was, but he enjoyed tricking us into committing what was tantamount to an ROCS crime: not preceding every response with, "Reserve Officer Candidate (fill in your name), yes sir!" He would get us laughing at one of his stories and then say, "Mister Sanders, what do you think of that?" Start your response without the proper introduction and he would say, "Ten demerits." Enough demerits meant no liberty.

One day he came into the classroom with a big grin and a newspaper folded under his arm. After class started, he held up the newspaper which featured, on the front page, a photograph of the collision of two Navy ships in the San Francisco Bay. His glee was unrestrained. It seems that the mishap—involving two Reserve destroyers, and would likely result in some ruined careers—was the funniest thing he had ever seen. My buddies and I did not share his mirth. We were in the same Navy—comprised of both active duty and reserve ships and personnel—right?

A favorite class of mine was one on weapons and in particular the 5-inch/38 caliber naval gun. We had to memorize all of its parts and functions. Little did I know that in a year's time—I would be in one of those gun mounts and firing at a foreign shore—in a country that I had never heard of. Of the people who were then making decisions that would affect my life—people like John F. Kennedy, and later Lyndon B. Johnson, Robert McNamara and Ho Chi Minh—I was familiar with only one. Not surprisingly, this was John F. Kennedy because I had watched a series of presidential debates on TV in 1960.

There was also a popular song titled *PT-109* by Jimmy Dean that came out in 1962, a portion of which follows:

> In forty-three they put to sea thirteen men and Kennedy,
> Aboard the *PT-109* to fight the brazen enemy.
> And off the isle of Olasana in the straits beyond the roo,
> A Jap destroyer in the night cut the 109 in two.
>
> Smoke and fire upon the sea—
> Everywhere they looked was the enemy…

The only danger we faced while at ROCS was from seagulls. One day it was my turn to be Section Leader of the Day. One of my duties was to march our company to the chow hall for meals and from one class to another. As we marched to and fro on base, so near the ocean, we were always at risk of having our uniforms splattered with a greasy, smelly substance falling from the sky. We ran "afoul" of one such bird that fortunately only hit the shoulders of three guys and not their heads. Practically every naval training base is near the water, so this phenomenon is pretty common to the experiences of every sailor.

One of the simple pleasures for young hungry sailors was "Geedunk." Every guy in the service knows the term, which dates back to at least World War II, and which refers to goodies such as candy bars, cupcakes and other snack foods. The Geedunk truck on base offered such delights, as well as drinks, sandwiches, and other fare that satisfied our desire to eat something besides Navy chow. Of course, our desire for snack food did not rate anywhere near our desire for entertainment and leisure in town.

While at ROCS, we were able to go into town in uniform on liberty, which provided a welcome break from the rigors of our academic/military training regimen. During one such foray, the Newport Folk Festival was taking place. There were many musicians in parks and along the streets, standing or sitting on blankets, singing and playing guitars or banjos. Being a rock & roll guy, I was not familiar with either of the featured musicians, Bob Dylan and Joan Baez. The presence of these future legends of Folk music were only pleasant distractions on my quest to find drinking establishments.

My buddies and I went to the Viking Hotel on Bellevue Avenue and started imbibing beer. The Viking Hotel was a very impressive and ornate building; we had nothing like it in Chico. I was only nineteen and not very worldly. I noticed a particularly engaging young lady and began putting the "Halldorson moves" on her. At some point, I began to wonder why she was smiling and laughing so easily at my corny jokes. She did look very familiar, could it be that she was from Chico? It suddenly struck me that I had seen her at my company commander's office a few weeks earlier. It was not long before she was joined by her husband, and I quickly lost that "Lovin' Feeling." Fortunately, I had not said or done anything discourteous or, at a minimum, I would probably have drawn extra duty back at the barracks.

On another liberty, I was enjoying a cold beer in a tavern. Sitting at the bar was a sailor wearing a uniform I did not recognize. I took a seat next to him, and struck up a conversation. It turned out that he was a sailor assigned to a German destroyer visiting Newport. He did his best to teach me some German. The only phrase that I picked up from him was "schone junge madchen," which he told me meant "beautiful young girl." This knowledge would come in very handy much later in my post-Navy life when I was an art student in Germany.

As the novelty of spending most of my liberty time at "watering holes" wore off, I sought to enjoy some of the attractions for which Newport is famous. One of these is a mansion built in 1895 for Cornelius Vanderbilt II, called The Breakers because the Atlantic surf breaks on the shoreline that comprises the boundary of the breathtaking grounds behind the mansion. There are a number of mansions along Ocean Drive in Newport, which was, and still is, a summer vacation spot of the rich and famous. The Breakers is the most spectacular of these estates. I was impressed with its seventy rooms and Italian Renaissance-style palazzo design inspired by the 16th century palaces of Genoa and Turin. Being an artist I was awestruck by some of the tapestries. The back lawn went on forever, or so I thought, with a gazebo out near the edge of the Atlantic. After visiting the mansion, several of us went down to the beach and swam in the relatively warm ocean waters.

One Saturday night, all of us, except those with too many demerits to be allowed off base, were transported to a fancy ball at a very ornate building with large windows from floor to ceiling. We were all in dress uniform and there, waiting for us, was a plethora of young ladies. They made quite a picture in their beautiful gowns. I danced with a particularly charming creature, but was not overly distracted because I had a girl back home, who would write to me almost daily and end every letter with, "I will love you forever."

Near the end of ROCS, we spent much time preparing for a "Pass in Review" on the parade grounds. Our company and the other companies were to march by a reviewing stand in which President Kennedy would be in attendance. All of the officer candidates looked sharp and performed flawlessly only to find out that the president's schedule had changed and he was not able to attend the event. I know, now, that his absence was prompted by the necessity to deal with an urgent matter in a troubled Southeast Asia.

4

Active Duty

Now is not the end. It is not even the beginning of the end. But it is, perhaps, the end of the beginning.
—Winston Churchill

Following Reserve Officer Training School, I returned in fall 1963 to my classes at Chico State College, and was having a great time majoring in art. I was walking to my Art History class when I heard the news. A high school classmate driving by in a Chevy Impala yelled out his window to me, "The president has been shot!" I thought that he was joking but, as I walked to my classroom, there were girls crying as they came down the hall toward me, which led me to believe that John Picot had been telling the truth. Upon arrival in the classroom, I found there was no professor present, only a note on the blackboard directing the students to go to Laxson Auditorium. There, a college official on stage stated that the president had been shot in Dallas, Texas. As I took in this news in disbelief, a faculty member walked over to the speaker and whispered something in the official's ear. We were then told, "The president is dead!" Our nation's innocence was stolen by Lee Harvey Oswald, who killed Kennedy with rifle fire from the 6th floor of the Texas School Book Depository, in Dallas, where Oswald was employed. Our beloved president was gone forever, and we were a nation in mourning and would remain so for some time.

Completely unrelated to this tragedy, I partied too much at Chico State that fall and received a letter from the college asking me not to return for a while. Girls, cars, alcohol, and a part-time job, on top of a full academic load had been my undoing. I had really enjoyed attending ROCS for the first of what was supposed to be two consecutive summers en route to becoming an officer when I graduated from college. Now with the news that I had "flunked out" of Chico State, and not wanting to commute one hundred miles each day to attend the nearest junior college, fifty miles away in Yuba City, I decided to go on active duty. I was already in the naval reserve, and had an obligation to serve two years on active duty as a part of my six-year enlistment.

TREASURE ISLAND, SAN FRANCISCO

> *Education is the ability to listen to almost anything without losing your temper or your self-confidence.*
> —Robert Frost

I was then a non-designated seaman in the naval reserve, but due to my high entry test scores was eligible for a variety of rates (jobs) offered by the Navy. An active duty sailor assigned to my reserve center in Chico suggested that I attend Electronics Technician Class "A" School at Treasure Island, San Francisco. He told me that ET "A" School was a very prestigious school and I did not want to enter the fleet as an ordinary seaman, which would entail my being assigned to the "deck force" aboard ship, chipping paint and doing other such menial work. Of course, sailors of all varieties do some type of menial work aboard ship. An example is a group of sailors, called a "working party," hauling heavy stores (provisions) from the pier up a steep brow (gang plank) to stow aboard one's ship.

My attendance at this school required a one-year extension of my active duty obligation. The course of instruction proved challenging and there was not any of the nurturing that is more common today. Almost on a daily basis, our senior enlisted instructor would stand in front of the classroom and bellow, "There is no doubt in my military mind that you are the biggest bunch of f**k-ups the Navy has ever produced." There were some variations on this theme, one of which was, "It is obvious to the most casual observer that you are the biggest bunch of f**k-ups in the whole U.S. Navy!" Whether he had distain for us, or was merely trying to toughen us up was not evident to me. Those who struggled with math, including me, were remanded to "Stupid Study" (Dilbert Study) for two weeks of intensive math instruction. One of my classmates and a good friend, Keith Girard, did not have to suffer Dilbert Study. Following his naval service as an ET, he worked as an electrician at a nuclear power plant in southern California. Mandatory study was, however, the Navy's way to try to ensure your success in a particular school. Flunk out and you would be going to a ship as an undesignated seaman and find yourself chipping paint!

A test was given at the end of the two weeks of Dilbert Study and if you did not pass, you endured two more weeks of grueling remedial math and tests. I took my final test in a second-floor classroom with no air-conditioning on a day when it was unseasonably hot in San Francisco. The stress was incredible for the poor bastards who were struggling to stay in ET "A" School. You needed to pass; you desperately needed to pass. The poor guy in the seat in front of me stressed so much that he had an epileptic fit. His arms and legs were straight out in front of him and he was choking. The instructor rushed over, assessed the situation, and called for help. The sailor was taken away and the test resumed. I too had my difficulties and did not pass the examination that would have returned me to the school curriculum. Ultimately, I would be sent to the fleet as a non-rated seaman. Similar to my experience at Chico State College, I had spent too much time enjoying myself versus fully applying myself to the task at hand.

While at the Treasure Island training command, I had a college buddy who lived in nearby Richmond, California. Gene Smithson and I had met at Chico State in the Fall semester of 1961 while taking industrial arts classes, and we both really liked fast cars and women. I had decided to buy Gene's '56 Chevy, and one Friday night we were working feverishly in his garage to get the car ready for me to test drive it home to Chico. We finished "wrenching" about midnight.

This beauty—which had more problems than were evident to me—was "cherry" with a bored out Corvette V-8 engine and a four-speed stick; for a young man of that era, cars just didn't get any better than that. It was metallic green with orange "nerf bars," metal devices which replaced the front bumper. The theory was that since nerf bars were much lighter than bumpers, their use increased the car's power-to-weight ratio, making it faster. The racer (our term for a hot rod) had four-inch headers that connected to dump tubes behind the front wheels—which gave the hot rod a menacing look—and exhaust pipes that led aft. The car was everything I dreamed of, in stark contrast to the 1956 four-door, 6-cylinder Chevy, with the shifter on the column, that I owned.

Gene's brother, Larry and his buddy were planning on going to Alaska with four dollars between them and wanted a ride north. Both had been rejected by the Marines. Apparently despite the tattoos of eagles they had acquired on their chests, other factors had failed to impress the Marine recruiters—likely a lack of good judgement. We rolled out of Richmond on the San Francisco Bay, around midnight, and headed for Chico, 180 miles to the north-northeast. By the time we reached Williams, a little farming town approximately one hundred miles up the road, the linkage for the first and reverse gears had gone out. I pulled into a Chevron station for repairs. As I exited the well-lit station, I did not realize that I had failed to turn my headlights on. The car's engine was roaring because "jury rigged" header gaskets (pieces of tin foil that Gene had installed) had blown out on the highway. Leaving the station, the fire-breathing hot rod sounded like an F-4 fighter aircraft preparing to take off. The engine had a three-quarter race camshaft—which allowed greater quantities of gas and air to enter the combustion chambers, producing more power and noise—and a muffler rendered ineffective by exhaust leaks from the header dump tube.

Photo 4-1

My '56 Chevy did not have the necessary pedigree to be turned into a hotrod; it was a lowly four-door family car, versus a two-door "screaming machine."

As I was waiting at an intersection for the traffic light to change, I noticed a police officer to my left, across the intersection, writing a ticket for a car he had pulled over. He noticed me and motioned with his flashlight for me to pull over as well. As I started to comply with this order, one of my two passengers said, "You can get away from him!" I don't know why I listened to a guy the Marines had rejected, but I did. After the light turned, I drove one block, turned left, then left again, then pulled over and turned off the engine. We were then one block west of our original position. We could see the police officer searching for us on perpendicular streets to the north and south. Fortunately, he never came down the street we were on.

After a while I crept out of town. When we got to Colusa, an even smaller farm town farther north, a sheriff pulled us over. I asked him what I had done, because I was going the speed limit. He replied, "There's a man from Williams who wants to see you!" I got as pale as a ghost and felt sick to my stomach. Within a short time, the Williams police officer I thought I had successfully evaded arrived and jumped out of his cruiser. I handed him my Navy ID card and he said, "You know Mike, I was in the Navy, the most I would have given you was a $12.00 ticket, but now I am going to get you for $300.00 if I get you for a penny."

He wrote me up for speeding, reckless driving, driving without lights, and eluding a police vehicle. I had not been speeding nor driving recklessly, but that didn't matter. Due to my poor judgment, I had to pay a large fine, leaving me insufficient money to buy Gene's car. After arriving in Chico, I also had to tell my father what had transpired. He took a day off from work, and travelled to the Williams courthouse for my hearing. Because he vouched for my good character, the judge lowered the fine from $300.00 to $200.00. This helped some, but still left me short of the "scratch" necessary to buy my dream car, leaving me with my "heap" and not Gene's cherry ride.

Problems with cars, and also girls and alcohol, would continue to haunt me. As discussed later, most of the problems with cars resulted from me spending all my spare time drinking and pursuing girls instead of doing maintenance on my cars. For readers much younger than I, who today own reliable cars, the automobiles of the 1950s required much work to keep them on the road. One Friday afternoon I left for Chico, and made it just past Fairfield when suddenly my ride developed a shaking that made me think that the engine was going to tear loose from the motor mounts. I had to pull over and have my car towed to a holding yard until I could figure out what to do about it. I then hitchhiked to Chico, because I had a date the following night with my girlfriend. She was a beauty and I did not want to disappoint her and, in the process, really disappoint me!

Saturday night found the two of us in my parents' car, parked in the back row of the Starlight Drive–In Theater. We were engaged in a tangled and heated display of affection, when all of a sudden there was a rapping on the driver-side window. I looked up at a man who I immediately recognized as her father. Was he going to pull me out of the car and thrash me within an inch of my life? He had always been so cool, and he liked me. (The way things had been progressing with his daughter, it's a good thing that he didn't show up ten minutes later). I rolled down the window and meekly said, "Hi." Mr. B. was some kind of a special guy. My girlfriend must have told him about my dilemma with my car, prompting him to come to the drive-in. He smiled and asked for the keys to my Chevy, as a prelude to going to Fairfield to rescue my car. He brought it back to Chico for me and parked it in front of my parents' house.

My car needed a lot of work, and I was not going to give up precious liberty time that could be spent drinking and dating to work on it. I offered to sell my car to Mr. B for one hundred dollars, thinking that he might want to fix it up for his daughter, who did not have a car at the time. He declined. He likely also did not want to devote all his time to working on it, and besides, the engine was probably blown. So, the car remained parked in front of my parents' house. My Dad eventually told me later, on another weekend liberty, that we should at least look under the hood. He opened the hood and removed the valve cover, and immediately discovered the obvious problem. A retainer holding the valve spring in place had sheared off allowing the valve to drop down and cause all of the problems.

Dad and I went down to the local junk yard, Jack's Auto Wrecking on South Park Avenue, with the retainer in hand. Jack had previously helped me many times with used parts for the '53 Chevy and '50 Olds coupe I had previously owned. Back then, working on your car was a weekly chore. We showed him the retainer and asked if he had something like it. He scuffed his boot through the gravel on the ground and picked up an exact duplicate of what we needed. We asked him how much, and with a grin he said, "Nothing." Now that's customer service. We returned home. Dad took the valve spring and put it in his table vice. He cranked it down and twisted a piece of wire to hold the spring compressed. We took the spring and the retainer out to the car. After installing it, Dad snipped the wire and the spring sprang into place. I left the car in Chico and eventually sold it. It was spotted around Chico for years, and may still be running today.

Not all my weekend liberties were spent in Chico. I also found time to goof off in the Bay Area. I was now assigned to the transient barracks at Treasure Island while awaiting orders to the Fleet. One night a couple of the guys and I decided to go into San Francisco and see *The Longest Day*, a war movie about World War II. The plan was to get some alcohol to put in our cokes, and thereby make the experience more memorable. Since we were all underage, we first needed to find a willing "angel" who would buy us booze. Perhaps we could find a drunk by a liquor store, and problem solved. Who could have guessed that angel would come in the form of a Hell's Angel, leathers and all. While we waited outside, our new-found friend got each of us a half pint of whiskey, for a nominal surcharge. Returning to the theater, we went into the show with our contraband tucked neatly in the waist band of our uniforms. Our plan, to drink and have a great time watching a super war movie, went awry, however. The theater's ice machine was broken, and therefore no ice. Drinking hard liquor in a warm coke made me sick, and I spent the remainder of the night "calling Ralph on the porcelain telephone" (heaving into a toilet).

Morning came early. I was still under the weather and unable to stand up—let alone report for muster. So I stayed in my bunk. Apparently, there was one hard-and-fast rule in the transit barracks, "Thou shalt not rest your weary hung-over ass on your bunk during working hours," which I had evidently missed in orientation. The barracks master-at-arms, a torpedo man's mate second, who was making his rounds, spotted me. He shouted some obscenities that included references to my heritage and various parts of my body. He then told me that because of my gross infraction, I was going "mess cooking" the Navy's equivalent to Army KP duty. I went over to the chow hall where, confronted by the smells of all manner of food, I turned green. The cooks took one look at me and promptly sent me to sick bay. A kindly doctor asked me if I had "made the beach" the night before. I told him that I had. He asked if I had eaten something that did not look good. When I told him I had not, he asked if I drank anything. I told him that I had, and explained how our plan had unraveled. He took pity on me and gave me a "twenty-four-hour no-duty chit."

The torpedo man's mate spied me back on my rack and came over to me screaming, "What are YOU doing here, you blankety-blank-blank-blank?" I held up my 24-hour no-duty chit, with a grin, and he went away. I suspected that he might try to get some measure of revenge the next day. However, fortune smiled on me that night; I received orders to the *Hopewell* (DD-681) stationed in San Diego. That night I called home ecstatic about being assigned to a destroyer.

5

Life Aboard the *Hopewell*

No man will be a sailor who has contrivance enough to get himself into a jail; for being in a ship is being in a jail, with the chance of being drowned. A man in a jail has more room, better food, and commonly better company.

—Samuel Johnson, in a letter written in 1759

Destroyermen have always been proud people. They have been the elite. They have to be proud people and they have to be specially selected, for destroyer life is a rugged one.

It takes stamina to stand up to the rigors of a tossing destroyer. It takes even more spiritual stamina to keep going with enthusiasm when you are tired and you feel that you and your ship are being used as a workhorse.

It is true that many people take destroyers for granted and that is all the more reason why the destroyer Captain can be proud of their accomplishment.

—Adm. Arleigh "31-knot" Burke, former commander Destroyer Squadron Twenty-one and the longest serving chief of Naval Operations

The first *Hopewell* (DD-181) was named for Pollard Hopewell, who died a hero in the War of 1812. Hopewell joined the Navy as a midshipman and on 21 August reported to the frigate *Chesapeake*. On 1 June 1813 the *Chesapeake* stood out of Boston Harbor and met the British frigate HMS *Shannon*. Disabled by two quick broadsides in the ensuing fifteen-minute engagement, *Chesapeake* struck her colors. Among her 146 casualties, both Captain James Lawrence and Midshipman Hopewell were killed. Captain Lawrence is best known for uttering the words "Don't give up the ship" as he was being carried below—mortally wounded by small arms fire. The second *Hopewell* (DD-681), my ship, was launched at the Bethlehem Steel Company yard in San Pedro, California, on 2 May 1943; sponsored by the wife of Adm. Raymond A. Spruance. The ship was commissioned five months later on 30 September 1943, Comdr. Corbin C. Shute, USN, in command.

Chapter 5

Photo 5-1

The destroyer *Hopewell* under way during the 1960s; she would be my home the next two-and-a-half years.
U.S. Naval History and Heritage Command Photo #NH 89658-KN

The Navy put me on a train to San Diego. I and the many other sailors aboard it were attired in our "blues" (wool uniforms) since Treasure Island was a rather cool climate. As we got closer to San Diego we were able to look out the windows and see scantily clad "beach bunnies" tanning themselves, guys surfing, and guys making out with girls. We so wanted to be on a southern California beach. When we got to San Diego, I was cut out of the herd and transported to Naval Station San Diego, located at 32nd Street and Harbor Drive. I decided to search the piers sequentially on foot, beginning with Pier 1, for the *Hopewell*. I hoped she would be in port and moored at one of the first piers. No such luck. The piers were long, it was the middle of June, hotter than blue blazes, and I was carrying a heavy sea bag. At Pier 5, I asked a sailor coming toward me where the *Hopewell* was. He said "Oh you mean the "Hopeless," she's over there." My new home—long, sleek and gray—was the outboard ship in a nest of five destroyers at Pier 6, moored outboard of the destroyer tender *Dixie* (AD-14). That meant that I had to cross five ships, stopping at each in turn, saluting the ensign (flag) and the officer of the deck, showing my identification, and requesting permission to cross—as well as making my way across numerous brows (gang planks), carrying a very heavy sea bag. The canvas bag contained Navy-issued regulation uniforms, some toiletries and of course the Blue Jackets Manual devoted to describing, for enlisted men, all the rules, regulations and minute details of Navy life.

Photo 5-2

Formal photograph of the author taken in 1964.

While I was at Treasure Island, the Navy had allowed me to request three types of ships on my "dream sheet." My first choice was a carrier, followed by a cruiser or a repair ship. I wanted to be on a BIG ship. You can imagine my surprise when they gave me orders to a destroyer, and an old one at that; the *Hopewell* had been commissioned before I was born. I wanted to be part of the crew of a big ship because of my belief that strength in numbers equated to more safety at sea. However, time would prove my great fortune in being assigned to a small ship where there was more esprit de corps and comradeship. Everyone knew your name, and you knew theirs. My shipmates and I were "tin can sailors," tin can being a fleet-shorthand for a destroyer. The nickname arose in World Wars I and II when sailors claimed their ships were made from tin cans because enemy gun rounds penetrated them so easily. The ship's designers had specified thinner hulls to enable the ships to achieve greater speeds. Due to their speed and maneuverability, duty aboard a sleek destroyer was considered glamorous.

HOPEWELL'S GUNNERY PROFICIENCY

> *The good ship USS* Targeteer, *"the world's smallest aircraft carrier," returned to port last week 700 pounds lighter than when she left. The weight loss was due to the gunnery excellence of* Hopewell *(DD-681). The Hopewell shot down two* Targeteer *drone aircraft, and thus became the second double drone killer in the* Targeteer *"Double Drone Killer Club." Even more significant,* Hopewell *is the first ship to ever shoot down both drones in one day during a single gunnery exercise.* Targeteer *operates out of San Diego as a unit of Service Squadron One, and provides target services to the fleet. The drone aircraft, or "birds," mission is to provide realistic anti-aircraft target practice for gun crews of ships undergoing fleet training.*
>
> —Article "Hopewell Becomes 2nd Double Drone Killer" from an early 1964 fleet newspaper (date unknown).

At the time I reported aboard in June 1964, the *Hopewell* had just completed a rigorous and lengthy pre-deployment training period in local waters off San Diego. On 5 May, a six-week extensive under way training program was begun during which new personnel were trained and old timers refreshed in gunnery, seamanship, engineering, tactical and battle evolutions. During gunnery exercises, the ship shot down two drone aircraft in two consecutive passes and was presented an award by the Drone Aircraft Catapult Control Ship USS *Targeteer* (YV-3). This unique vessel was a former World War II era Medium Landing Ship. With refreshed training completed on 19 June, we began a leave-and-upkeep period in preparation for a forthcoming WestPac (Western Pacific) deployment. On 1 July, the *Hopewell* transferred from Destroyer Division 52 to Destroyer Division 172.

RITES OF PASSAGE

To any newly reporting young sailor, the first few days aboard a Navy ship are, at best, confusing and, at worst, scary. For me, coming from a small town in northern California, there were a number of culture shocks awaiting me in the Navy. Among them, living in close quarters or associating with a diversity of people from all across the social spectrum and from all across America, as well as from other countries such as the Philippines and Guam. (Filipinos, Guamanians, Samoans, Puerto Ricans and other men from Allied countries and U.S. territories could enlist in the Navy and earn U.S. citizenship following a certain number of years of service.) You had to learn where everything was, and it was all foreign to you. Where was the chow line? Which side of the ship did you walk forward on and which side did you walk aft? Where was Sickbay? Where did you sleep, and could you find the space again, deep in the bowels of the ship? Were you assigned a General Quarter's (battle) station and, if so, where was it?

These were only a few of the questions I had. Most of my new "best friends" were friendly and helpful; others not so much. Newly reported sailors or "boots" were an irritant to some of the guys who reminded us that they had worn out more sea bags than we had socks. The division to which you were assigned normally provided an "old hand" termed a "sea daddy," who was supposed to introduce you to some of the basics. However this person might or might not be enthusiastic about the additional duty thrust upon him. Ideally, the sailor was top notch, and not a black sheep seeking to introduce a new man to their view of the Navy; but not always. Eventually things settled down and after some passage of time, you became one of the "old hands" to newcomers and had opportunity to show them the ropes. Remembering my first days aboard ship, I always tried to be as helpful as I could to new guys.

I did not realize that when partaking of meals aboard the *Hopewell*, you ate only with your own kind: other sailors assigned to your division or department. Being new to the "mess decks" (the area where enlisted men dined), I did not know this protocol. I sat down at the "snipes" table, and immediately became the focal point of many

Chapter 5

pairs of eyes. The engineers aboard ship are called snipes because they work in "the hole"—the very hot, very noisy machinery spaces below deck.

At this table was a boiler tender first class, hell bent on feeding himself as fast as he could. Up to this point in my life, I had not seen a guy put his head down and shovel food into his mouth with two forks—one in each hand. I don't think that guy could have afforded his food bill in the civilian world; he had to stay in the Navy to survive.

I later had occasion to see this same, rather large BT1 "on the beach" (in town) in San Diego. A shipmate and I had gone to the movies and we ran across him, attired in his Hong Kong custom-made suit, going in at the same time. The first class boiler tender had two large tubs of popcorn cradled in his arms and two large cokes, one in each hand. He looked very happy. We probably should have asked him to join us, but were mindful that he was a snipe, and therefore scary to us, but nicely dressed. Sailors visiting Hong Kong typically bought a suit there. Hong Kong tailors were legendary for their skill and speed in which they could turn high quality fabrics into a perfect suit. When I later made my first visit to Hong Kong, I was no exception. I got measured and bought a suit that fit perfectly, as well as accompanying shirts monogrammed with my initials. Over the pocket and on the cuffs were "MRH" for the whole world to see that these fine threads belonged to Michael R. Halldorson. I was stylin'. I even bought some pointed "Beatles" shoes. What are you waiting for girls? Here I am!

The BT1 was not the only interesting person with whom I dined aboard the *Hopewell*. Every ship had a stud—a cut-up, too cool for school—who also might be feared or revered, or both. A signalman third filled this role aboard the *Hopewell*. Everyone called him "Tate." My first encounter with Tate was at chow. With my food tray in hand, I chose to sit at a table that just happened to include Tate. He must have thought, "Here's fresh meat," because he looked at me and started to grunt. As his eyes bulged and he stared at me like a madman, his grunts became louder and louder, and their frequency increased. The grunts were menacing. Suddenly his hand went into his mountain of spaghetti, pulled out a fist full, grunted again and, thrust the wad of spaghetti into his face. Looking at this seemingly deranged "shipmate" with spaghetti all over his face and smeared down the front of his dungarees, I made a hasty retreat to the entry doorway, looked back, and then left the mess decks without eating. Based on this episode and others, I was sure that Tate would be sent to a Naval Hospital for observation and released from active duty "for the good of the Naval Service." However, he continued to serve his country and also served to amuse for the remainder of his time aboard the *Hopewell*.

We also had other colorful characters aboard. One such was a boiler technician named Bowman. During his spare time, which shipboard engineers normally had little of, he liked to fish for sharks. When he caught one, he would pickle the eyes and save one or more of the teeth, strung on a leather thong necklace which he had to give evidence of his prowess as a shark catcher. While on deployment the uniform of the day in hot climate was dungaree pants with white tee shirt. You could always tell it was Bowman walking toward you as his white tee shirt typically had dried blood on it from new teeth cutting him from beneath it. Now there was a tough guy; some might believe one who needed his head examined.

One spectacle I did not particularly enjoy was viewing the tattoos sported by some of my shipmates. Anyone who has taken a public shower in the military knows what I am referring to; a plethora of so-called art work, displayed on bodies sometimes equally unattractive. It was like being trapped in the comic section of the Sunday newspaper. Most artful-appearing tattoos were acquired in Japan or Hong Kong; others were done aboard ship by God-only-knows who, and by what method. The proverbial heart with a dagger through it and the name of a wife or sweetheart were common, as were these names later obliterated by ink. More imaginative ones included hinges on the owners elbows, pigs screwing on top of a sailor's feet, spider webs almost everywhere; a wrist watch that never changed time, and a tattoo across a belly asserting "beware of swinging boom!"

Some of my shipmates decided that I needed a tattoo and said that they would get me ashore, get me drunk and have a tattoo placed strategically. They also told me that they were going to make me smoke cigarettes. I replied that, should this happen, "I will hunt you down and kill you the next day." They did not pursue this plan, likely because they realized that I truly did not want my body decorated with their idea of art and I was not a smoker. My threat of doing them bodily harm was probably not a deterrent; rather they were mindful of my wishes and respected them.

FIRST UNDER WAY

My first time "under way" involved the *Hopewell* "breasting out" to allow another destroyer to join the "nest" of ships of which she was a part, moored outboard of the destroyer tender *Dixie* (AD-14). Ships undergoing repairs or maintenance by *Dixie* needed to be alongside the tender. Since we were not having any work done, we needed to move clear of her, out into the channel, and then reform the nest. Despite San Diego's harbor being tranquil that day, and the *Hopewell* only rolling gently back and forth as we moved out into the channel, I became seasick. From then on, well before getting under way from a berth or anchorage, I would take Dramamine. If I forgot it, I paid for it.

I was also to learn that having fun at the new guy's expense continued at sea. I was assigned to "mail buoy watch," which involved standing on the main deck at the bow of the ship, exposed to wind and sea spray, while watching for the mail buoy—that little bobbing beacon of hope that would facilitate many guys on board receiving word from home. I remember thinking, for a moment, how truly amazing it was that the Navy could know where to station that buoy so that our ship would pass that exact point in the Pacific. But, I didn't fall for that one, and thus did not take up the "mail buoy watch," although for a split-second I was in the "I want to please" mode. Other men new to the ship were asked to go down into the hole to ask the snipes for some "relative bearing grease," or to bend over and peer into a cardboard box in order to view a "sea bat" contained within. As one sailor was swatted, as was customary, with a broom, he exclaimed "stop hitting me, I am trying to see the sea bat." Did I ever fall for any of these type pranks that the old salts aboard relished so much? The answer, my friend, is blowin' in the wind.

ASSIGNMENT TO THE DECK FORCE

> *I never saw quite so wretched an example of what a sea-faring life can do; but to a degree, I know it is the same with them all; they are all knocked about, and exposed to every climate, and every weather, till they are not fit to be seen. It is a pity they are not knocked on the head at once, before they reach Admiral Baldwin's age.*
>
> —Jane Austen, *Persuasion*. This quote pertains to the day of sail, but, the seamen that do ship's work topside in today's navy also are often worn out at a fairly early age.

Being a non-rated seaman, I was assigned to the deck force. A non-rated seaman was a person who had not been sent to, or had not completed A-School before arrival on board a ship. Once a part of deck force, it might be possible for particularly hardworking sailors who desired to work in another area of the ship, to leave the deck force. However, because their transfer elsewhere aboard would leave the deck force short one man, their division officer had to really believe in their potential in order to give them up. Some of the men that stayed in deck force, most of the non-rated seamen, became rated boatswain's mates. Thus, the deck force (formally designated 1st Division), was comprised of a few boatswain's mate petty officers and many seamen, over whom they exercised supervisory control (power). Responsibilities of the men of the deck force included such things as:

- preservation of the exterior of the ship
- rigging for evolutions such as refueling and anchoring the ship
- standing bridge and lookout watches, and operating boats

Since Navy ships are continuously exposed to salt water spray, continuous "rust busting"—chipping off old paint and painting over bare spots with a primer of "red lead," followed by a coat of Navy deck or haze gray—is required. The old saying, "once over dust and twice over rust," is only a joke, particularly aboard *Hopewell*, which was old and had two wars under her belt; a seaman assigned to chip paint on the fantail, had actually broken through the steel deck with his chipping hammer. The deck force was also responsible for maintenance of the deckhouse and superstructure, as well as the decks and sides of the ship. On one occasion, I was hoisted many feet up in the air to paint the forward engine room stack "stack black." The purpose of this dull black paint was to disguise soot that adhered to the top of the stack or, as known aboard British ships, the funnel.

Of course, seamen, whether rated or non-rated, did whatever they were told to do aboard ship. One of my more unpleasant tasks was assignment to a "working party" tasked to remove sacks of rotting potatoes from the 01-level. Sailors eat a lot of potatoes, and they were commonly stored topside, on some ships in a bin, but aboard the *Hopewell* out in the open. After many sacks of rotten potatoes, each weighing about thirty pounds, my dungarees stank to high heaven and I could not get them off soon enough.

During my stint in the deck force, I started drawing cartoons—either because I fancied myself an artist, or to relieve the monotony of "busting rust"—which provided me a bit of status as "the ship's artist." My favorite cartoon was a stylized version of a Don Martin face on a skinny body, the face with the corn cob smile showing rows of large teeth. (Martin was an artist for the satiric *Mad Magazine*.) My work usually consisted of a basketball player with a string bow-tied to his finger and attached to a basketball. For some reason I thought this was funny and so did many of my shipmates. One day, a member of the crew asked me if I could draw the same character but with large lumps on his neck. Well of course I could, and I did. I did not inquire as to why the drawing was desired; I just wanted to please.

A few days later, a different crewman came up to me and asked how I could be so cruel. I looked at his neck and seeing large bumps there, I realized I'd been had. I explained that I had not seen him aboard before, and that when I was asked to draw the caricature, I had not known the reason for the request. I felt about two feet tall. He then understood what that crewman had asked me to do and that I meant him no harm. We were steadfast friends from then on. Some people can be very unkind, for whatever reason. My guess is that it makes them feel better about themselves.

IN PORT DUTY

While the *Hopewell* was in port in San Diego I stood duty like everybody else. Aboard Navy ships, there is a requirement for sufficient officers and men to remain aboard each day to get the ship under way in an emergency. Aboard the *Hopewell* we were in three-section duty. Every third day, one-third of the crew was aboard for the entire 24-hour period. One of my duties was to patrol the 01 level with a carbine on my shoulder to deter any boarders. One night as I was making my rounds, the ship next to us was showing topside on the 01 level the nightly movie for the enjoyment of men now on watch or otherwise engaged. I paused to see what the movie was. It was *From Russia with Love* starring Sean Connery, and was very good. Up until then I had not seen a James Bond movie. I paused for a while and took in a good segment of the movie before I remembered that I was supposed to be guarding my ship.

LOCAL OPERATIONS

> *Sleeping on watch is a serious offense against the UCMJ [Uniform Code of Military Justice]. More than that it is an abandoning of responsibly to your shipmates. Anyone found sleeping on watch or otherwise inattentive to duty will be severely dealt with.*
>
> *Speed of action is important in battle but foolhardy carelessness can be as dangerous as the enemy.... Remember your ammo handling safety precautions. Mount captains are responsible for training and drilling their men.... Men not assigned to weapons stations are cautioned not to skylark around such lethal weapons.*

—Excerpt from a memorandum by the *Hopewell*'s executive officer, titled Information and Preparations for the Tonkin Gulf Patrol, of September 1964.

Prior to leaving on a deployment, a Navy ship spends much time at sea conducting training and local operations to prepare for, and demonstrate proficiency in, its primary missions. During this period, I stood lookout watches, and when the ship was at General Quarters (battle stations), I was assigned to a gun mount. In accordance with the nautical rules of the road, every ship is required to have at least one person assigned to dedicated lookout duties. The caveat "and they must be human" was added after a merchant ship was involved in a collision, and the court of inquiry found that the master was employing the ship's mascot, a dog, for lookout duties. Aboard a Navy ship, there are typically at least two individuals assigned such responsibilities, termed the forward and aft lookouts. The forward lookout is normally stationed aboard the bridge and the aft lookout on the fantail. The latter individual is very important, as the officers and other watchstanders on the bridge cannot see the area behind the ship. Navy lookouts, which helped safeguard the ship during combat operations, were trained through the use of silhouettes of enemy aircraft, ships and submarines. Because I was a "boot" (newly arrived on board) and was going to be assigned to "mess cooking," I received next to no training.

One night at sea, while standing lookout watch topside, I sighted through my binoculars red and green lights well above the horizon coming toward us. There was no mistaking the running lights of an aircraft coming at us, but still at a distance. I dutifully reported an aircraft inbound. Why hadn't CIC (the combat information center) picked up this aircraft on radar and what if it was hostile? I was pretty proud of myself for detecting the intruder until an old hand informed me in a kind manner that the aircraft I was reporting was Mars. It seems that due to atmospheric conditions, the red planet can at times appear to be both red and green. Who would have thought?

I was a little better in fulfilling my duties as the second loader on Mount 32, one of the ship's three 3-inch/50-caliber gun mounts. My job was to carry projectiles from bulkhead storage to the holder on the side of a twin 3-inch mount, so the first loader could feed them into the hopper to be cycled into the gun. The first loader was "Big Jake"—all five-foot-two of him. He was a very good first loader, always giving his all. Everything was relatively simple as long as the gun was elevated for anti-aircraft fire. When the twin-barrels were raised, the hopper in the breech was much lower, and was thus easier to load. When we went to surface action, it was another story. The hopper went higher into the air as the barrel lowered to engage a target on the water's surface. Big Jake would stand on his tippy-toes endeavoring to feed 3-inch rounds into the gun. In one instance he did not quite make it. The round hit at his feet, bounced off the steel deck, and rolled toward my feet. Startled and a little sick to my stomach, I picked the round up, and pitched it over the starboard side of the ship. It was a "point detonation" round, and my first thought was that it might detonate and blast shrapnel all over the deck, killing the gun crew.

My second GQ station was in Mount 51—the forward most of the *Hopewell*'s four 5-inch gun mounts, two located forward and two aft—as a fuse setter. My job was to operate the equipment that set the fuse time on projectiles with mechanical fuses. I monitored the Fuse Indicator Regulator, and selected the fuse setting ordered by matching the setting sent from the main battery director. I sat way down under the gun, watching the "crow's feet" indicators from the main battery director rotate either to the left or right to indicate the fuse settings. From that location, the pungent sulfur smell of gunpowder, extremely loud boom-boom-boom of the gun firing, and the violent shaking of the mount were not pleasant. Moreover, in rough seas, the ship pitched under this mount. On one occasion the safety officer, positioned above me inside the mount, vomited on my head. Thank goodness I was wearing my ball cap.

MESS COOKING

After a few weeks of deck force "knuckle busting," I was greatly relieved to be assigned to mess cooking. Mess cooks assist the cooks (commissary men) by performing all forms of menial tasks associated with feeding the crew and cleaning up after meals. One morning, as I was making toast in a rotating conveyor toaster across from the serving line, I was accosted by a rather large fire plug-shaped individual named Ryan, but commonly called "Tiny." He elbowed me away from the toaster and, after I re-gained my footing, I asked him just what the hell he thought he was doing. He said, "Mind your paygrade lad!" knowing that I was just a lowly seaman. Later that morning, I had occasion to venture up to the fo'c'sle (the forward part of the ship) and saw Ryan working on a gun mount. I asked someone if he was a third class gunner's mate. They told me he was not, that he was a gunner's mate striker which meant a seaman, the same pay grade as me. A striker is the term used to indicate that he was studying to be a gunner's mate.

Following a repeat performance by him the next morning at the toaster, I cautioned him, "Mind your pay grade lad!" He did not like that. He looked at me with his steely gray-blue eyes, the kind of eyes that go right through you. He replied, "Not now, but sometime today, I'm going to push your nose through the back of your head!" Later that morning, I was walking up the port side of the main deck just as Ryan was coming down the same side. We met in "no-man's-land" where no one could see the two of us from the forward or aft parts of the ship. And to make things worse, the 01 level above overhung the main deck at that particular spot, so no one could see us from above the main deck either.

We were virtually invisible to everyone. What were the chances of this particular set of circumstances happening? Pretty slim, I would image. Oh God, why me? It seemed an appropriate time to mentally recite the 23rd Psalm. What was that passage? "Yea, though I walk through the valley of the shadow of death, I will fear no evil: for thou art with me; thy rod and thy staff they comfort me." Ryan stared at me and I stared at him. Then I said the bravest thing I had ever uttered in my short life, "I'm not afraid of you!" That, of course, was a bald-faced lie. My knees were shaking. For some inexplicable reason, he took no action but instead replied, "I'm not afraid of you either." We cautiously moved around each other, counter clockwise I think, and then we gingerly moved past each other and remarkably, my nose was still in place and the air never seemed so fresh.

After that incident, Ryan appeared to like me to some degree (put up with me would probably be more accurate). Unfortunately for me, he was not always nice. A year or so later, we were refueling alongside an oiler at sea when the ships suddenly moved apart. The refueling hose broke loose and sprayed the deck, side and topside area with black fuel oil. The seamen on the refueling detail were ordered to swab (mop) up the oily mess. I was, by then, a petty officer and as such it was my duty to order Ryan to swab up the oil around Mount 53, one of the after gun mounts located centerline on the 01 level. Ryan did not respond to authority particularly well. He tried to break the handle of the swab he was using across my shins. Yee-ouch that hurt! I did not report that incident because I did not need the grief that he would visit upon me if he really got mad.

During a bar brawl at Subic Bay, Philippines, that took place while the *Hopewell* was on a deployment to the Western Pacific, reportedly eight shore patrolmen hit Ryan in the head with night sticks, and he still put several of them in the hospital.

PREPARATION FOR DEPLOYMENT

> *In the last two months three patrols have been made [from the South China Sea] into the Tonkin Gulf. All have been attacked by PT boats: the Maddox on 2 August, the Maddox and Turner Joy on 4 August, and the Edwards and Morton on 18 September. Our patrol will be number four. Let's be ready to defend against any attack.*
>
> *Knowledge of your job at GQ is the best insurance against getting overly excited in battle. The men who have been shot at and missed are usually those who remain cool and efficient in tight situations. Don't panic if the torpedoes start to whiz by. Just do your job and trust that others will do theirs.*
>
> —Excerpt from a memorandum by the *Hopewell*'s executive officer, titled Information and Preparations for the Tonkin Gulf Patrol, of September 1964.

About this time, the *Hopewell*'s crew generally became very excited about our pending cruise to the Western Pacific, which officers and enlisted men alike referred to as "Westpac." There was no war going on and the "old salts" aboard told the "new hands" that drinks in the bars were cheap, and there would be girls readily available to us. A few men did not share the euphoria of the others because they were married or did not want to leave a girlfriend in San Diego. One of the ship's first scheduled tasks after arriving in Westpac was to patrol the Taiwan Strait as a deterrent force to help protect Taiwan from mainland China. A shipmate, and good friend, Jim Brickey, who had experienced this activity during a previous cruise, told me that we would visit Kaohsiung, Taiwan, where the most beautiful women in the world resided. My response was, "What about California girls?" They were then well-known to the world, being touted by the Beach Boys in a popular song.

Photo 5-3

Servicemen during the Vietnam War dreamed of girls back in the States, just as had men who served in earlier wars.

The Gulf of Tonkin Incident changed our tasking, however. It began with an exchange of shots on 2 August 1964 between the U.S. destroyer *Maddox* (DD-731) and North Vietnamese torpedo boats in the Gulf of Tonkin. The *Hopewell* would now proceed to Vietnam by way of Subic Bay, Philippine Islands, versus the Taiwan Strait. After leaving San Diego, we would make the normal stops in Pearl Harbor, Hawaii, and Subic Bay for food, fuel, maintenance, and crew liberty, and then sailed for "Yankee Station" off South Vietnam. Yankee Station was a point off the coast from which carrier-launched aircraft flew strikes against North Vietnam. We would "plane guard" for carriers, which involved steaming behind an aircraft carrier in order to retrieve any aviators that went "in the drink" due to equipment failure or pilot error during the launching or landing of their planes. Much more of our time was devoted to providing naval gunfire support to ground forces ashore. Shore bombardment involved providing protective defensive fire to help safeguard friendly forces by keeping the enemy away, or putting "hot lead" on the enemy. Hence, I never got to Taiwan.

OUTBOUND TO SEA

After months of pre-deployment preparations, the *Hopewell* stood out of San Diego on 5 August with other units of Cruiser Destroyer Flotilla Eleven, bound for the Western Pacific where she would join the mighty Seventh Fleet. We arrived at Pearl Harbor on 12 August and Midway Atoll, much farther west, on 16 August. On 23 August we were in Guam and eight days later, we arrived at Subic Bay Naval Station, Philippine Islands. While there, Comdr. Donald F. Milligan, USN, relieved Comdr. Robert A. Moore, USN, as commanding officer. From there our destroyer division was sent to relieve the *Turner Joy* and *Maddox* and two other "tin cans." Now, instead of the joy of a peace-time cruise, there was the realization that we were at war. I now had some concerns for my own safety as well as that of my ship. Pangs of dread came over me at times. There were so many ways to be hurt or even lose your life aboard ship. One could be washed overboard and lost at sea, or injured or killed by a shipboard fire or an explosion. Now, there was also the possibility of hostile action. North Vietnamese PT boats and MIGs (fighter aircraft acquired from the Soviet Union) were operating in the area and we might find ourselves their target.

The term "he lost his life" (or she lost her life) has always sounded a little foreign to me. How does one lose their life? Did they misplace it? He died or he was killed would be more accurate. Would I be KIA (killed in action) or wounded? Although these thoughts entered my mind, I did not dwell on them. Daily shipboard life was busy and all consuming.

6

Initial Portion of the Deployment

> *We clear the harbor and the wind catches her sails and my beautiful ship leans over ever so gracefully, and her elegant bow cuts cleanly into the increasing chop of the waves. I take a deep breath and my chest expands and my heart starts thumping so strongly I fear the others might see it beat through the cloth of my jacket. I face the wind and my lips peel back from my teeth in a grin of pure joy.*
>
> —L.A. Meyer, *Under the Jolly Roger: Being an Account of the Further Nautical Adventures of Jacky Faber.* This quote adequately describes the excitement I felt upon standing out of San Diego, headed west, on my first deployment.

During the first part of the cruise, I was still assigned as a mess cook, had developed a fondness for cooking, and decided to become a commissary man, Navy terminology for a cook. I needed to cook, it was my calling. More importantly, if I were to become a cook, I would escape the deck force. I did not want to return to the deck force and become a boatswain's mate. The other option was to become a "snipe" and work in the innards of the ship where it was as hot as Hades; temperatures in the fire room could reach 130 degrees. Snipes ate salt tablets like candy, and sweat so much they looked like they had just stepped out of the shower fully clothed when they emerged from machinery spaces. (Such an occurrence was rare, non-designated seamen normally stayed in deck force or were assigned to "above deck" jobs. However, it did occasionally happen if needs of the ship prevailed.)

Thus, I was striking to be a commissary man, studying the Petty Officer Third Class Manual in preparation for a test which, if I passed, would result in my being advanced to CS3. The cooks wanted me to join their ranks because I was the only one who could figure out how many 25-pound cases of dry stores would fit into a vestibule. (Vestibules served as transition zones between compartments on the same deck, and if they contained ladders, between decks. Every free space aboard a ship was packed with provisions before it left home port on a deployment.) It was simple math. Everything was going along so well and I knew the information contained in the third class manual backwards and forwards. No returning to the deck force, and I would not have to become a snipe. Life was good. Life was really good!

The *Hopewell* was a pretty good feeder, and the cooks tried as much as possible to provide choices. For example, when they served fish, "once in a blue-moon," they also offered hotdogs for those who did not like, or could not eat, fish. However, the cooks did engage in some guile from time to time. A little trick that I learned from them was to crack a few real eggs and stir them into the powdered variety to make the crew think their entire serving of scrambled eggs was real. A piece of shell was a stroke of genius.

No matter how diligent the cooks were about cleanliness, occasionally you would see a cockroach dash across the mess decks. The roaches came aboard as larva in cardboard boxes of canned goods and dry stores. Cockroaches can subsist on almost anything, including the glue in the cardboard boxes, and an oily fingerprint. Smaller ships often discarded boxes on the pier and brought aboard only the cans; a ship the size of the *Hopewell* did not have this option. The ship's corpsman sprayed pesticides on a regular basis to kill them, but they were hardy. Such problems are not unique to American ships. In fact British sailors call their master-at-arms aboard ship "crushers" because as the "Navy policemen" patrol their ships at night, their big feet crush cockroaches caught unaware. Sprayed roaches retreat into voids behind false bulkheads on the mess decks, where their carcasses stack up until "rip out work," done during shipyard periods, uncovers them.

When the *Hopewell* was later in a shipyard overhaul following deployment and the mess decks were being refurbished, the lagging (insulation) over steam lines that traversed the space was cut away for replacement. There were hundreds of cockroaches—live ones as well as carcasses—that spilled out and had to be dealt with. The pipe lagging, used to protect crewmen from being burned, was made of asbestos. We did not yet know the linkage between this type material and the lung disease Mesothelioma. Johnny Sharp had one of the top bunks in the OC (operations and communications) compartment, and had to sweep asbestos flakes off his sheets with his hand each time he went to bed. Deteriorated lagging-covered pipes were above his head. When a more senior person from OC Division transferred off the ship, Johnny took a lower rack. That was a good move.

PRESSED INTO YEOMAN DUTIES

One night at sea, a group of cooks, a few ship's servicemen and I were sitting on the main deck outside the galley door. A ship's serviceman first told us about his duty aboard a carrier before being assigned to the *Hopewell*. He had us spellbound with an account of a jet pilot who, when launched, did not have the necessary air speed to clear the front of the carrier. (Such incidents are normally due to a casualty to the catapult resulting in inadequate steam pressure to fling an aircraft skyward.) Knowing that he was going into the "drink," in a burst of inspiration, he decided to use the plane's after-burners to propel it deep enough that the carrier's propellers would not shred both it and him. The SH1's audience was greatly impressed by this story. Later after becoming more seasoned, I realized that the pilot would have had insufficient time to benefit from afterburners before his plane hit the water. And, if he had, the plane would have likely broken apart upon impact. Hanging out with ship's servicemen and commissary men made me want to become one of them.

My dreams of becoming a cook were thwarted, however, by the XO, the acronym for executive officer, the officer second in command aboard a U.S. Navy ship. One day as he entered the mess decks to conduct his daily morning inspection, I stood at attention and saluted him smartly. I might have said, "Seaman Halldorson standing by for your inspection, sir." Sailors aboard ship soon learned that if you stated "standing by, ready for your inspection," there would be more things found for the person presenting the space to correct. He said, "Halldorson I hear you are striking to be a commissary man." I replied with an emphatic "Yes, sir." He said, "No, you're not. I didn't approve a request 'chit' (submitted by someone on my behalf) for you!" Talk about your whole world collapsing. Here I was, getting ready to take the third class commissary man test, and now I would be going back to the deck force after my tenure as a mess cook. A deck ape again! I had better sharpen my seaman's knife, learn more knots, and be prepared to stand a lot more lookout and bridge watches, including mid-watches.

Have you ever stood a mid-watch, from midnight to 0400 (four a.m.)? Well I have, and so have scores of sailors before me, and there will be scores after me. Your watch starts when you are rudely awakened by a seaman assigned as "the messenger of the watch" or, if not bound to the bridge, someone from another watch section aboard ship, such as radio central, CIC (the combat information center), or engineering. You then go down to the chow line, with sleep in your eyes, for some "mid-rats" (midnight rations). This meal is intended to fortify oncoming watch standers to help them stay alert during their watch, and provide fare to the off-going watch before they hit their rack. Alert and awake is a GOOD thing when you are on watch. However, because ships are only allocated enough money to feed enlisted men three meals a day, mid-rats are typically leftovers. This does not normally present a problem unless, for example, you are eating greasy sausage when the seas are particularly rough—which might mean the food comes up faster than it went down.

Later that day, the XO walked up to me on the fantail, where I was sunning myself and feeling very sorry for myself prior to my reassignment to the deck force. He said, "Halldorson, I see that you have had some typing and two-and-a-half years of college; I need a yeoman." Without hesitation I stood up and said, "I'm your man!" Suddenly life was good again. As it turned out, I really enjoyed paperwork, even more than cooking.

LIBERTY IN SUBIC BAY, OLONGAPO, PHILIPPINES

I turned twenty-one while the *Hopewell* was at Subic Bay en route to Vietnam. Ships on deployment frequently visited Subic Bay, because in addition to its naval station, it also boasted a naval shipyard capable of effecting almost any kind of needed repair or overhaul of machinery. I had duty the night of my birthday and could not "go over" into town to celebrate. In fact, I don't believe that I "made the beach," for whatever reason, during that stay. The next time we visited Subic, no one was allowed to leave the naval station because the heads of four decapitated sailors had been found floating in "Sh*t River" wrapped in a sheet. This waterway—which flowed, or oozed, between the naval station and Olongapo, an adjacent city—was termed thus by sailors due to the fecal matter it contained. Whether these men were murdered in conjunction with a robbery or as an act of retribution for something they did, I cannot say. This shocking event highlighted that it was prudent to be careful when ashore overseas.

The first time I got to go on liberty in Olongapo was with Jim Brickey, a friend and "old hand." He took me to the Orchid Club with which he had much previous experience. Being from Chico, and not particularly worldly, I asked him if we could "get a date" with a couple of the bar girls. How was I supposed to know that a date involved the transfer of money? He must have thought "what a dolt," but arranged our "paying two girls' way out of the bar." (This involved compensating the overseer of the girls, for the loss in revenue their leaving with us would incur.)

Map 6-1

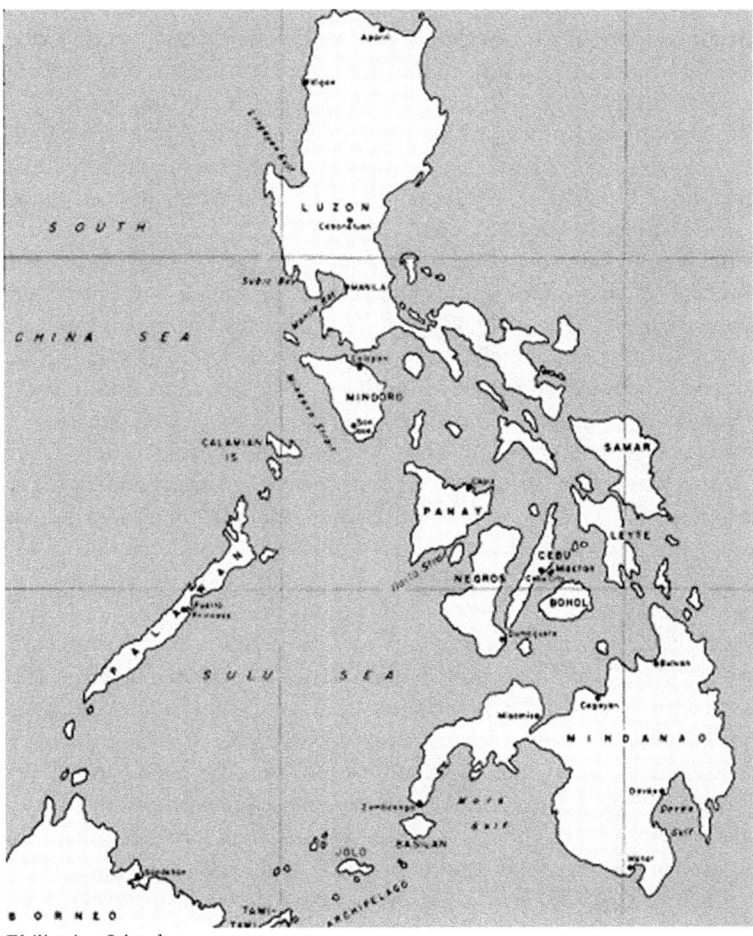

Philippine Islands

We went to a nearby bowling alley and started drinking San Miguel beer. My date, Esther, started saying "sorry about that," an expression that was "all the rage" in 1964. She must have said it about a hundred times, until I just wanted to throttle her! So while Jim and the two girls were bowling, I slipped out the back door and prowled the streets until a cute little number named Anita picked me up. I fell in "love" at first sight. I kept telling her that she looked like my girlfriend back home. Eventually, she took me to the bar where she worked, the Orchid Club! As we walked in the front door, I spotted Esther at the bar and she saw me. (Apparently, she had quit the bowling alley after I had departed.) Anita had to use the restroom so I sat down at a table. Esther came over and began yelling some unpleasantries about my heritage and threw her drink in my face. (Although I did not know it at the time, my actions were apparently those of a "butterfly boy," a patron that flitted from girl to girl.) I started laughing, and she slapped me so hard that I fell out of my chair. I laughed even harder, not so much at her as at the situation, but she did not know that. Incensed, Esther went to the bartender and asked for her "butterfly knife."

The Navy does not always fill you in on all local customs. It seems that bar girls in the Philippines claim you if you are with them for any length of time. If you then go to another girl, you have "butterflied" on them, which tends to make them mad. The girls have a butterfly knife for self-defense and—if they get mad enough—to teach customers a lesson. This type knife, also known as a balisong or fan knife, was a folding pocket knife with two handles such that, when closed, the blade was concealed within grooves in the handles. The bartender gave her knife to her and Esther was advancing toward me as Anita came out of the head. Anita quickly deduced what was going on and threatened to kill Esther if she tried to harm me, thereby perhaps saving my life.

Preceding one of my ship's visits to Subic, we learned that Motown's The Supremes were scheduled to perform at the EM (Enlisted Men's) club on Naval Station Subic Bay, and we would be in port at the time of their concert. I knew the name of the lead singer, Diana Ross, and loved their latest song, "Where Did Our Love Go?" Then, the Supremes cancelled their engagement. Diana, where did our love go? On my last liberty before the *Hopewell* proceeded south-southwest to Vietnam, I went to the EM club and started drinking with "Little Doc," a term used by sailors for the junior enlisted "corpsman" (hospitalman) aboard their ship. Little Doc and I were feeling no pain when he returned aboard ship in the late afternoon to give VD shots; I remained at the EM Club and continued to drink. A bunch of cooks from aboard my ship sat down with me and we were having a good old time until we began playing a drinking game called "Do you want to fight?" One guy would say, "Do you want to fight?" and the other guy would say, "Yeah." Both would stand up (if they could), put up their dukes, fake a punch, and then laughter would break out. The "warring" parties would then pat each other on the back or shake hands, sit down and continue drinking. That is not what happened on this occasion.

It appears that the cooks held a grudge toward me for becoming a yeoman instead of one of them. When the Shore Patrol (members of the crew performing "military police" duties ashore) found me, I was in the bushes outside of the EM club. They returned me to the *Hopewell*. The next day we deployed for Vietnam. When I awoke that morning, my mouth was cut inside and half of my left front tooth was sheared off. To make matters worse, I was hung over, did not take Dramamine before getting under way as was my usual practice, and became violently seasick.

BAD NEWS FOLLOWS INJURY

At mail call later that day, I got one letter and it was from my girlfriend back in Chico. Before I opened the letter, I said to myself that I would write her back and tell her that this letter was the best medicine I could have gotten for all my ailments. The letter started off, "Dear Mike, Ralph and I have been going steady for a month now." This was the same girl who had written me at Officers Candidate School the summer before and had ended all of her letters with "I will love you forever!" Apparently forever had a short shelf-life. When you first embark on your Navy adventure, you expect to lose a few things as a result of your Mom cleaning house and throwing away your prized baseball card and *Mad Magazine* collections, and hope that she didn't find the *Playboy* under the mattress. One thing you don't anticipate is losing your girl to some suave, sweet talkin' guy with a hot car.

7

Latter Part of Deployment/Combat Duty

> *The fact we are going on the patrol in the [Tonkin] Gulf [off South Vietnam] is still classified information. Do not write home about it and possibly alert our adversaries.*
>
> —Excerpt from a memorandum by the *Hopewell*'s executive officer, titled Information and Preparations for the Tonkin Gulf Patrol, of September 1964.

The *Hopewell* was administratively under the command of commander Cruiser-Destroyer Force, U.S. Pacific Fleet when in home waters, and a part of the mighty 7th Fleet when deployed. She was a unit of Cruiser-Destroyer Flotilla 11, Destroyer Squadron 17, and one of four ships in Destroyer Division 172. Every Navy ship had a unique call sign which they used when communicating by radio with other military units; during the Vietnam War the *Hopewell*'s call sign was "Allegheny."

The *Hopewell* was a 2,100-ton *Fletcher*-class destroyer, named for Admiral Frank F. Fletcher. *Fletcher*s, like all destroyers, were armed with various anti-ship, anti-aircraft, and anti-submarine weapon systems. While I was aboard her, *Hopewell* had four five-inch gun mounts and an additional three 3-inch/50-caliber mounts. The 5-inch/ 38-caliber guns were the mainstay of U.S. Naval surface combatants. They were capable of firing 55-pound projectiles a maximum distance of fifteen miles, but nine miles was considered its effective range. The maximum range was how far the projectile would fling the gun round, whereas the effective range denoted the distance at which you could reasonably hope to hit your target. Our guns were very accurate due to both their design and the men who fired them.

Hopewell also still had hedgehogs, an antiquated World War II-era anti-submarine weapon which employed explosives fired from racks on the port and starboard sides of the ship just below the bridge. Their 7.5 pounds of TNT made quite a bang. Hedgehogs were developed by the Royal Navy to supplement much larger depth charges rolled off the fantail of a ship. These much smaller mortar bombs exploded on contact and helped make the enemy more cautious about attacking a warship or a group of merchant vessels shepherded by one. Interestingly, we had none of the much more desirable torpedoes for use against enemy submarines; they had been removed to free up space for installation of the 3-inch guns. Because the 3-inch guns could be trained and elevated more rapidly and also had a greater rate of fire than the 5-inch guns, they were preferable for use against aircraft. The 5-inch guns were best for shore bombardment and use against ships if such a requirement arose.

EXPOSURE TO "BROWN SHOE" DUTY

The *Hopewell* left Subic Bay the morning after my face met too many fists. Thus, I was unable to be attended to by the base dentist ashore. Fortunately, a helicopter from the carrier we were operating with was dispatched to pick me up and take me to the *Bon Homme Richard* (CV-31). As I stood on the fantail of *Hopewell* waiting for the harness to be lowered from the helicopter hovering overhead, a shipmate who was helping me get ready instructed me to bend my legs. He explained this action was necessary to absorb the shock if the ship pitched under me while the aircraft took a strain on the cable, else I could break a leg. I took his advice.

Once aboard the carrier, the dentist told me that they only had one shot of Novocain for me and would, therefore, have to pull the stump of my broken tooth without the benefit of additional painkiller. I told him I could take it, and take it I did. He made me a plastic partial to replace my tooth.

I was not able to immediately return to my ship, so, while aboard the carrier, I was able to watch night flight operations. Aircraft launching and returning with flaming jet exhaust piercing the darkness was quite inspiring. While observing flight ops during the day, I would look over at the destroyers, mine included, whose duty was to screen the carrier from submarine, surface and air threats. One moment you could see all of the "tin cans" and the next their bows would disappear into a wave crest. In contrast, the massive "bird farm" was barely moved about by wind and wave. I also learned that carrier sailors ate well, pretty much anytime they wanted. Aboard the *Hopewell* the serving line was open only at breakfast, lunch, dinner, and for mid-rats. If you were hungry between meals and were lucky, you might catch the ship's store open and purchase a candy bar. Aboard the *Bon Homme Richard*—named after John Paul Jones' ship lost off England—

hot meals were available twenty-four hours a day on one or more of the ship's many mess decks.

When it was time to leave the carrier, I was catapulted off in a COD (Carrier Onboard Delivery) twin-engine, propeller-driven plane used to ferry personnel, mail, supplies, and high priority cargo, from shore to ship and ship to shore. Following the "cat shot," which involved accelerating from zero to two hundred miles an hour in seconds, the aircraft flew to Cubi Point Naval Air Station in the Philippines. I was temporarily assigned to the naval air station while awaiting the return of my ship from sea. At this point, readers probably recognize that in the eyes of Navy leadership, it was less important to devote aviation resources to return me to my ship than it had been to provide me emergency dental care. While at the naval air station, I learned that aviators based there also ate well. My ship eventually arrived in port and I left the 'brown shoe navy" to return to the "black shoe navy," the distinction being that aviators wore brown shoes.

Having lost my girlfriend, I decided to seek more dangerous duty, believing that I should thus relieve someone else in harm's way who still had a girlfriend, or more importantly a wife. Of course, I had not yet had much sea time, let alone seen any combat action. Still simmering with anger and feeling hurt over the "Dear John" letter I'd received, I requested to be transferred to "Swift Boat" (fast patrol craft or PCF) duty off the coast and in some inshore waters of the Republic of Vietnam. The executive officer denied this request and I was very upset with him. He alleviated my anger, to a degree, by explaining that it would take him six months to train another yeoman to do as good a job as I did in the Ship's Office. Many years later, I thanked the XO for denying my request and possibly saving my life.

CATALYST FOR THE UNITED STATES' FORMAL ENTRY INTO THE VIETNAM WAR

> *Too long we have fixed our eyes on traditional military needs, on armies prepared to cross borders, on missiles poised for flight. Now it should be clear that this is no longer enough—that our security may be lost piece by piece, country by country, without the firing of a single missile or the crossing of a single border. We intend to profit from this lesson. We intend to reexamine and reorient our forces of all kinds—our tactics and our institutions here in this community.*

—Statement made by President John F. Kennedy in a speech to the American Society of Newspaper Editors on 20 April 1961.

In late September 1964, the *Hopewell* and other ships of her destroyer division relieved the destroyers *Maddox* (DD-731), *Morton* (DD-948), *Richard S. Edwards* (DD-950), and *Turner Joy* (DD-951). *Turner Joy* and *Maddox* were involved in the Tonkin Gulf incident which led to the escalation of the war after Congress passed the Tonkin Gulf Resolution. On 2 August 1964, the *Maddox* was conducting surveillance of North Vietnamese coastal waters in support of American/South Vietnamese clandestine fast patrol boat (PTF) operations against North Vietnam, when it came under attack by three North Vietnamese P-4 torpedo boats. The collection of intelligence, termed "DeSoto" patrols, resulted from a desire by commander, U.S. Military Assistance Command Vietnam to obtain information on enemy coastal ground forces, and naval craft capable of intercepting South Vietnamese operated PTFs and PCFs (Swift boats) during missions and to identify radar sites that could detect and track the boats. The attack was repulsed by gunfire from the *Maddox*, which sank the torpedo boats. Following this incident, the North Vietnamese government claimed that its units had been chasing South Vietnamese PT boats, and had assumed that the destroyer was a part of the raid. However, a subsequent attack on the *Maddox* and *Turner Joy* the night of 4-5 August clearly appeared to be no accident as neither of the ships, which had been under surveillance by the North Vietnamese, was engaged in any hostile action. The Johnson administration quickly proposed, and Congress passed on 7 August, the Tonkin Gulf Resolution, which stated in part:

> ...the United States is, therefore, prepared, as the President determines, to take all necessary steps, including the use of armed force, to assist any member or protocol state of the Southeast Asia Collective Defense Treaty requesting assistance in defense of its freedom.

This legislation would serve as the legal basis for the armed military support provided by the United States to the Republic of Vietnam throughout the war.

Map 7-1

Indochina

Although I did not know it at the time, *Hopewell* also apparently conducted "DeSoto" patrols during her duty off Vietnam. The following are excerpts of correspondence I received from a shipmate who was involved with these operations:

> The Desoto patrols were during the 64/65 tour.... We welcomed on board several new personnel (I can't remember where they came aboard, maybe the Philippines). They were dressed in khakis with no markings (stripes or bars), and they remained on board for the duration of the patrols and lived in officers' country. They

assisted in installing additional equipment in CIC [combat information center]; two cameras were installed on the surface radar repeaters. We were on high alert for high speed torpedo boat radar contacts.... We had equipment to record the frequency, pulse width, etc. of emissions in the areas of the DeSoto patrol. All of this information was sent to the Pacific Fleet Communications Elint [Electronic Intelligence] Center in Japan for review. I believe that we sometimes did erratic ship maneuvers in hopes of the North Vietnamese energizing specialized equipment. We were always accompanied by another destroyer on these patrols.

ON THE GUNLINE OFF THE REPUBLIC OF VIETNAM

> *Port and starboard condition watches will be maintained while on patrol. Three mounts will be ready to fire on a moment's notice. They may be firing when GQ is sounded.*
>
> *All hands topside should be alert to report mines, torpedo wakes or other objects in the water. Don't rely on the two lookouts to see everything. Sing out loud and clear if you see a mine or torpedo. Other men pick up the warning and pass it along to the bridge by relay. It could make the difference between turning away in time and being hit. Torpedoes have a phosphorescent wake at night or can be seen at several hundred yards.*
>
> —Excerpt from a memorandum by *Hopewell*'s executive officer, titled Information and Preparations for the Tonkin Gulf Patrol, of September 1964.

After several weeks of training in the vicinity of the Philippines, *Hopewell* finally set a course west for Vietnam. One bright, clear morning in the South China Sea after arriving off the coast of South Vietnam, we were assigned a shore bombardment mission that involved firing into the jungle over a group of sampans between *Hopewell* and the shore. The Vietnamese fishermen were out in force that day, and there appeared to be hundreds of the small craft. We opened up with all four 5-inch guns and within minutes, the craft had all disappeared. The following morning, we returned to the same location and resumed fire. The sampans stayed right where they were and continued to fish, apparently having determined that they were not our targets and that we presented no danger to them.

Photo 7-1

Hopewell firing three of her four 5-inch/38-caliber mounts.

I was then part of the gun crew assigned to Mount 52, located just forward of and below the bridge on the 01 level. One muggy night, I was outside of the mount getting some shut-eye with the other guys (unless we were firing, we were only required to be at our station), when, all of a sudden, I awoke to see what appeared to be a pair of running lights approaching fast. My mind immediately leapt to the possibility of an enemy patrol boat. Were we going to get raked with machine gun fire or have a torpedo fired at us? And, how did this vessel get so close without us detecting and firing upon it? The other guys were asleep, and I was hesitant to wake them unless there was a real danger to the ship. Not to worry, as I rubbed my eyes gaining my night vision, the possible enemy combatant turned into a harmless fishing trawler, passing nearby in the darkness. Sailors new to the combat zone off South Vietnam often perceived danger in nearly everything they saw. Being "wound too tight" is not good; on the other hand, letting one's guard down could result in dire consequences.

When we first arrived in the Tonkin Gulf, and were briefed on operating procedures, we were told the North Vietnamese had patrol boats and that we might be attacked by them at any time. Accordingly, we trained rigorously with South Vietnamese fast patrol craft (PCFs), which mimicked the tactics that might be used against us. We initially found it hard to track these boats maneuvering at high speeds and bring a 5-inch gun to bear on them, but after some time we were up to the task. I was then assigned to the main battery director for the forward guns (located above the bridge), which provided fire control solutions to the two forward guns. Someone told me that my life would be short-lived if we were attacked by an aircraft, as my director and also the one for the after guns would be obvious targets. Shortly after I was very happy, when I was assigned to a gun mount. I did not stop to consider that enemy forces generally shoot at the things shooting at them!

We also trained with A-1 Skyraider attack aircraft in the South China Sea off South Vietnam, to polish up our anti-aircraft prowess. One day, a flight of these type planes was to practice bombing the *Hopewell* by dropping bags of flour, while we tracked them and simulated taking them under fire. During the ensuing mock attack, the planes did not drop these "dummy bombs," but I still found the experience unsettling. The aircraft came in from starboard and flew very low directly over the ship, way too low for my taste. Destroyers are crammed full of fuel and ammunition from stem to stern, and I was acutely aware that I was aboard a floating powder keg. Suddenly, being a sailor was not as inviting as when the Navy recruiter impressed upon me that being on a ship at sea was infinitely better than being in a foxhole during hostilities.

One night we were steaming a mile or so off the South Vietnamese coast searching for the pilot from a downed A-5 Vigilante. The Vigilante was a high altitude, carrier-based strategic aircraft capable of carrying nuclear weapons. Unfortunately we found neither the aviator nor the plane which he had ejected from or rode into the sea. We did find a shoulder harness, the tip of a wing, and a few maps floating on the surface, which could have been from another downed aircraft. This episode impressed upon me that flying off a ship off Vietnam was more dangerous than being aboard a ship off Vietnam.

Pilots who ejected from their aircraft and went down in the sea off Vietnam faced a variety of dangers, including drowning, hypothermia, and poisonous sea snakes. During the cruise, one member of the crew of the ship's motor whaleboat gaffed a sea snake as the boat was approaching *Hopewell*'s side and held it up so those of us topside could see. Even from a distance away, the snake appeared incredibly ugly. Pilots were afraid to be in the water off the Vietnamese coast because of the threat these snakes presented. If bitten and not treated, the snake's venom was reportedly fatal within twenty-four hours. The serpents generally traveled in swarms, which meant that the danger of being bitten was greatly multiplied.

During the near-shore search, our searchlights stabbing at the ocean's surface could have served as beacons to enemy shore batteries, patrol boats, or aircraft. Likely, if such a threat had existed, the searchlights would not have been on. Fortunately we were not fired upon from either the shore or the sea. The Viet Cong sometimes sent individual soldiers in small baskets out into coastal waters to shoot futilely at passing ships. (The best they could have done was to perhaps injure or kill one or more sailors on deck. A rifle could not stop a ship.) A shipmate on watch reported seeing the remnants of such a basket. We had altered course, and run over it and its crew of one. The officer of the deck must have been unaware of this report, as no action was taken to stop the ship and retrieve the basket and enemy personnel.

LIFE AT SEA

> *I spent uncounted hours sitting at the bow looking at the water and the sky, studying each wave, different from the last, seeing how it caught the light, the air, the wind; watching patterns, the sweep of it all, and letting it take me.*
>
> —Gary Paulsen, *Caught by the Sea*

When your ship is out to sea, operating independently—not steaming in a formation with other ships—all you see is water, sky and the horizon. If it is really foggy during the day or particularly dark at night with no moon, there might only be gray or blackness around you. It is quite an alone feeling. The vastness of the sea in comparison to a very small ship is humbling, and accents such feelings. Particularly while on deployment, I longed for letters from family, friends, and my girlfriend. It means a great deal to a serviceman to get word from home. My best friend, Robert Bruce Wadlington, attended Shasta Junior College in Redding, California, a small city about seventy miles north of Chico. Many people called him "RB," but I affectionately addressed him as Arb. He sent me humorous updates on home life in the States, some written while he was in class. He would take ads from *Playboy* and other magazines and insert outrageous lines which made me double over with laughter. As might be expected, my parents wrote me about more mundane things, which, while comforting, didn't provide the same comic relief.

I both wrote and drew on the envelopes of the letters I sent home. Following is one example from my second deployment:

Chapter 7

Photo 7-2

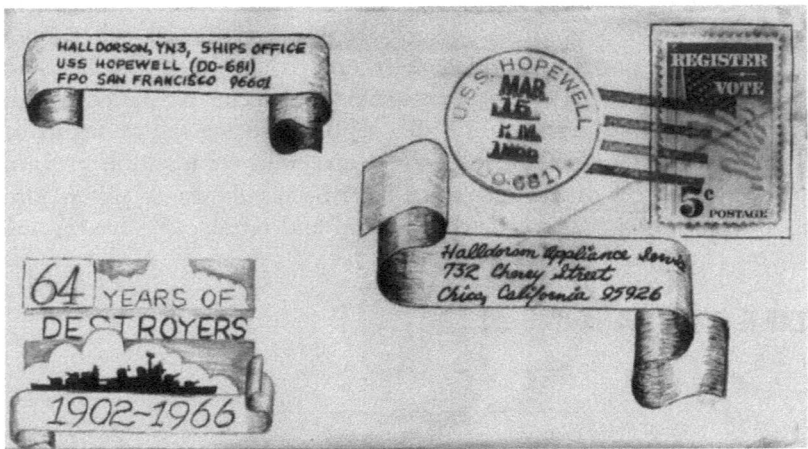

Photograph of an envelope displaying my artwork.

At sea, when not working or engaged in special evolutions such as refueling and firing the guns, there was occasionally some time for relaxation in the form of movies, visiting with other guys, or just time alone which some people devoted to reading books. I did not do much of that. Most of my time was spent drawing, but I did read two books that I can remember. The first one was a rather raunchy little paperback titled *Candy*, which made its way around the ship among the crew about three times. The second was titled *Diamonds are Forever* by an author I had never heard of, named Ian Fleming. I could not put the book down. I read it in the chow line and on successive evenings until I had finished it.

Following one of our tours of combat duty off South Vietnam, *Hopewell* visited Yokosuka for crew liberty. Tokyo, which is located about fifty miles north of Yokosuka, had hosted the 1964 summer Olympic Games some months prior. Some of *Hopewell*'s crew had the chance to visit Tokyo. The athletes being long-gone, the city was filled with tourists, vacationers and Navy types. The weather was great and we had a good time touring the Olympic venue and other Tokyo highlights. At the end of this pleasant interlude, my buddy and I boarded a train for the return trip to Yokosuka. Unfortunately we did not speak or read Japanese. Following departure, we saw a lot of the Japanese countryside, and as the miles rolled by, we did not seem to be getting any closer to a population center. We gradually began to feel that something might be wrong, but could not communicate our apprehension or ask for guidance.

An old man eventually took pity on us and informed us that we were heading west across Honshu and not toward the southern side of the island. As soon as we could, we changed trains and made our way back to the ship. On our errant ride, we met a very nice man and saw a lot of the countryside that we ordinarily would not have experienced. This episode highlighted that the world is filled with friendly, helpful people.

Photo 7-3

Dungaree jacket patches

While in Yokosuka, crewmen assigned to the Ship's Office, including myself, had patches made for our dungaree working jackets. Normally, men assigned to sea duty exhibit their ship's patch, but we wanted our own. I designed it, depicting a yeoman holding a quill astride a garbage can propelled by an outboard motor, our idea of a "destroyer yeoman." The motor bears the number "681" which is the hull number of our beloved *Hopewell*. At the top, I placed, "Far East Cruise," Navy terminology for duty with the Seventh Fleet. Another patch, commonly available on the street for anyone to purchase, was the "Tonkin Gulf Yacht Club." This patch, in yellow, red and black, depicted a Chinese junk. With the advent of the Vietnam War, more deployments were moved farther south, and such cruises came to be known as WestPac (Western Pacific) deployments.

BACK ON THE GUNLINE

Returning to the gunline off the coast of Vietnam, the change in weather from that of Yokosuka was dramatic. It was hot and muggy and time spent inside gun mounts—where there was little air movement—was particularly toilsome. Hour after hour was spent at General Quarters either firing or waiting for the order to "commence fire" again. We were in battle dress—shirt collar turned up with top button buttoned, and pant legs tucked into socks—preventing our dungarees from breathing. Unlike during most shipboard activities, there was little idle chat and jokes; for the most part guys just endured the conditions. Then would come the firing order, and we would all come alert and resume our jobs: I trained the gun to the correct bearing as indicated by the main battery director; the pointer elevated or depressed the barrel; the powder man placed the powder in the tray; and the projectile man released the projectile from the hoist clamps, placed it in the tray ahead of the powder, then inserted the projectile and powder into the gun. We were then ready to fire.

We were a finely honed crew and we could perform this operation in a matter of seconds. We had reason to be proud. But, we had little knowledge of our overall mission. We were not told details about a particular target, such as where or what it was. We did not know the effectiveness of our fire, or results of any battle damage assessment. Our reality was the smell of gunpowder, the noise and violent shaking of the gunmount, and the acrid air inside the mount. Occasionally the executive officer would publish the results of our shore bombardment in the Plan of the Day. We did know that we were supporting our guys inland and that our job was necessary and much needed to save American lives. Thoughts of what our troops were going through inland were very sobering. We suffered mind-numbing hours on station, accompanied by thirst and hunger, but such inconveniences were nothing compared to the combat ashore.

HONG KONG

Following her relief on station on 14 December 1964, *Hopewell* left Vietnam coastal waters bound for Hong Kong to enjoy the Christmas holidays in the exotic city, arriving there two days later. Hong Kong means "fragrant harbor" in Chinese. It got its name from an area on the south side of the island where trade for fragrant wood and incense products once flourished. Victoria Harbor, where Navy ships normally anchored and used liberty launches to convey crewmembers ashore, separated Hong Kong from the Kowloon Peninsula. From the deck of a ship at anchor the view of the city at night was spectacular; lit up like a neon rainbow casting light on the harbor surface.

When we pulled into Hong Kong, the boatswain's mates and seamen of the deck force were ecstatic because of the expectation of a dramatically reduced workload in port, thanks to Mary Soo and her girls. In exchange for brass shell casings from spent gun rounds, a group of women would give the sides and superstructure of a Navy ship a new coat of haze-gray paint. Following naval gunfire, a gun mount crew was expected to "police" (collect) all brass 5-inch shell casings ejected from the breech onto the main deck. They were to be retained and transferred to a Navy ammunition ship for return to the United States for reuse. Of course, sailors aboard some ships threw many shell casings over the side into the South China Sea, made artifacts such as ashtrays out of them, or saved them for barter in Hong Kong. Aboard *Hopewell*, we followed standard procedures and thus had no extra brass. Instead we traded leftover food for the desired work.

Photo 7-4

Canisters holding expended brass shell casings.
(Courtesy of R. Mike Sohikian, RD3)

While ashore in Hong Kong, the "old salts" aboard Navy ships usually got "liberty cuffs," silk panels stitched inside the cuffs of their dress blue uniforms. Even though I was not an old hand, I did so also. The Navy enlisted dress blue uniform was stylish, comprised of a jumper and bell-bottomed trousers topped off with a white hat. The jumper had piping all around the flap and on the sleeves, and the bell-bottoms had a thirteen button fly. Many sailors would contend the latter was not particularly easy to use, but it did look good. The lining of my uniform sleeve cuffs, like many others, featured ornate dragons in orange, red and green. Because they were not sanctioned by the

Navy, I would turn the cuffs outward to display my silk art only in bars where there were no officers present.

Many civilians in America also liked to wear the Navy jumper; it was particularly popular among female college coeds in the 70's, and is still beloved today. Used jumpers can be found at thrift stores—a zipper put in the front allows them to double as jackets. Aboard ship, a sailor's working uniform was dungarees: denim shirt and pants, and a zip-up blue working jacket. Thus, American sailors were commonly collectively referred to as "blue jackets" by their officers.

Some sailors went all the way and had custom "blues" made in tailor shops in Hong Kong. They were made of gabardine. Occasionally the Navy considered making uniform changes, despite the fact that the existing uniforms easily identified American sailors and we liked them. Once while in Hong Kong, the scuttlebutt aboard ship was that first and second petty officers were going to be authorized to wear khaki colored chief-type uniforms. One of our second class petty officers, a radar man, went ashore and got measured and fitted for his new "glad rags." He suggested that I also obtain some. He evidently thought that I would be a second class, soon. I did not pursue his suggestion. He ordered several sets and—you guessed it—the Navy changed its mind and did not authorize the uniform change.

Another common practice in Navy liberty ports was for local families to host one or more sailors for a home cooked meal. An English family invited a small number of crewmen from our ship to Christmas dinner. I was one of three chosen to go. Our hosts lived on Victoria Peak with a stunning view of Repulse Bay. To get there we rode a tram up the side of the mountain, enjoying the view all the way up. We were treated to a superb dinner, plus conversation with their pretty daughter made for a great time. I was excited to find that our host and hostess had every Beatle album released up to that time. Being British, they were very proud of the "Fab Four," and played some of their music for us. I loved the Beatles, and still do, and I was in rock & roll heaven. The dinner conversation inevitably turned to the Vietnam War. They were interested in what we were doing, and expressed their support of the United States' efforts.

The following day a chief aboard my ship asked me if I wanted to go have Christmas dinner with an American family. He had been invited but had a touch of illness and was unable to go. Thus, 1964 found me eating two successive Christmas dinners. The second affair was nice and pleasant, but not as interesting as the one with the English family. During these two evenings, I reflected on how, in

1944 my father had spent Christmas in Stalag IX-B, located southeast of the town of Bad Orb in Hesse, Germany. A diet of grass and water eventually reduced his weight from 150 to about 70 pounds. How fortunate I was, but my holiday was not indicative of all Vietnam veterans. Five years later in 1969, my brother Alan's ship was stuck on a sandbar near Vung Tau, Vietnam. In addition to this indignity and the possibility of being shot at by the Viet Cong while grounded, there was no hot meal for him or anyone else aboard.

Photo 7-5

A hand drawn postcard that I sent home to my family for Christmas in 1964.

While in Hong Kong, I shopped for a variety of items at plush shops at very good prices. I bought a particularly nice gold ring with an inlaid tiger's eye and wore that ring for many years, until I lost it. Later in the cruise, I bought cameras and a plethora of electronic gadgets at a reasonable cost. These included a state-of-the-art Akai reel-to-reel tape recorder, and a new Yashica camera with which I took many pictures. I was a fervent "shutterbug" and snapped anything and everything. During a subsequent visit to Hong Kong, I set up the camera on a tripod on the fantail of *Hopewell*. I set the exposure time for one minute and pointed it at the city. The results did not disappoint, although there were some lines resulting from movement of the ship under the camera. I had captured on film one of the most beautiful cities in the world, with night lights dancing in colorful arcs—something I had previously only enjoyed in person.

Of course, there were other much less wholesome activities that sailors could pursue while on liberty. In an effort to deter *Hopewell* sailors from engaging in one of these, there had been a note in the ship's Plan of the Day, warning us of the perils of seeking sexual pleasures on the roof tops of Hong Kong. Unfortunately, the Navy assumed that we knew that "rooftop" was a euphemism for any floor above the ground level, meaning "do not go upstairs with a prostitute." A more succinct explanation would have been "in many establishments, activities that take place above the ground floor result in VD," which could have saved a great deal of pain for those who partook of services offered by the ladies of the night.

On one run ashore in Hong Kong, I and a group of shipmates began our liberty in a drinking establishment at 10:00 a.m. Chinese bar girls came over to our table to talk and try to entice us into buying them copious drinks. Having limited funds, we weren't going to buy drinks for anybody besides ourselves. The girl who chose me was a tall, gorgeous Eurasian. She wore a Suzie Wong dress with a slit up the side. She was very sexy and tried her best, but still no drinks were forthcoming. She eventually tired of me and left to try her luck with another sailor. We got out of there and continued our pub crawl through Hong Kong.

In Hong Kong, most British and Australian sailors hung out at the China Fleet Club, their equivalent to our EM (Enlisted Man's) clubs. American sailors were also allowed. Unfortunately, a very few American sailors did not respect the hospitality afforded them by their hosts at this club, and engaged in a practice potentially dangerous to their health. These dolts would stick their heads in the front door of the China Fleet Club, yell "F**k the Queen," and then run as fast as they could, because, right behind them, would be the pride of the Royal Navy. A fun and more mature endeavor was to trade hats with Brit or Aussie sailors. They liked our white hats and we liked their "Donald Duck hats." The tricky part was trying to get back aboard your ship without the proper "cover," in both navies, I would imagine.

One of my fondest memories of Hong Kong occurred later in the cruise. While ashore on liberty, a typhoon passed through the area and, because of the dirty sea, we were unable to get back to our ship. The water launches that had transported us from the *Hopewell*'s anchorage to the city could not operate in those conditions. We were "forced" to stay at the Hong Kong Hilton that night, and several of us had dinner at the Eagles Nest, a restaurant on the 25th floor of the lavish hotel. My dinner of Kobe beef and shoestring potatoes cost about $3.00 American, the exchange rate being eighteen cents

American to a Hong Kong dollar. The meat was so tender you could cut it with your fork, and very tasty. I was told that this was because Kobe beef cattle were given a bottle of Asahi beer a day to fatten them up.

BACK ON THE GUNLINE

> *The usual delays in preparing to fire exercises when safety is paramount are unacceptable while on patrol. When General Quarters is passed hit the deck running and be prepared for anything when you arrive on GQ station. Mount crews must be ready to open fire IMMEDIATELY.*
>
> *When going to GQ stations remember the proper routes to reduce confusion and avoid traffic jams: Forward and up on the starboard side; down and aft on the port side. If General Quarters is sounded at night do not show any lights while going to stations.*
>
> —Excerpt from a memorandum by *Hopewell*'s executive officer, Information and Preparations for the Tonkin Gulf Patrol.

Hopewell departed Hong Kong on 28 December 1964 to return to naval gunfire support duty in the South China Sea. My third GQ station was "hot-case man" in Mount 53, a rear facing 5-inch gun mount. As sizzling shell casings were ejected from the rear of the gun, my job was to knock them down through an opening in the deck of the mount. To protect my hands, I wore large asbestos gloves. Besides being very hot, the five inch-diameter casings were essentially fast-moving sections of pipe which, left to their own devices, could inflict injury to personnel inside the gun mount. There were also other ways to be hurt inside a mount. On one occasion, we were firing ten rounds of rapid fire at a target with the gun in local control. In this mode, the gun was fired by the pointer in the gun mount (and not by the main battery director), and the mount crew had control of the gun.

We were in a big hurry to demonstrate how fast we could shoot ten rounds. Through his hatch in the top of the mount, the mount captain was observing where our shells were hitting. Suddenly, he crouched down and yelled "Cease Fire!!!" to the gun crew. Thank God he did! The projectile man had removed the round from the hoist and had thrown it into the tray with unusual haste. The correct way to position the projectile onto the tray was to hold it down with your fist to ensure that it stayed down and was rammed into the gun correctly. Our guy had not held this particular round in place and, as he turned hurriedly to retrieve another projectile, it bounced up into the air and came to rest in the tray backwards. But for the attentiveness of the mount captain, and his quick action in yelling cease fire, we could have had a mount detonation (if it had been rammed into the gun), killing us all.

Diagram 7-2

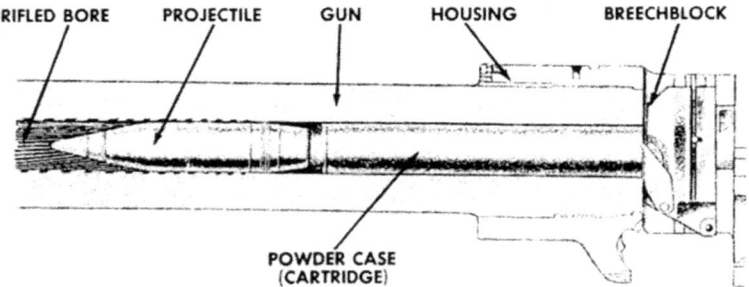

The *Hopewell* used semi-fixed rounds; the projectile and powder were loaded separately prior to firing the gun.

It is not unusual for sailors to be assigned new duties; cross training helps to ensure a well-trained crew. Thus, eventually my GQ station was "trainer" in Mount 53. The trainer slews the gun barrel(s) left or right to align, or train, the mount with the bearing of the target. I took great pride in my work, and endeavored to help put "hot lead" on the target as soon as possible. By closely watching the "crow's feet" on the instrument panel in front of me, I would immediately know which way the main battery director had ordered the mount to train. I would then crank brass hand wheels as hard as I could to manually position the gun. Once I had matched the ordered position, I would throw a lever to "auto" and the main battery director would take control of the gun's movement.

Required actions at one's GQ station were preceded by manning your station and preparing for battle. If you were in the after part of the ship when the General Quarters alarm sounded and your GQ station was forward—you ran up the starboard side of the ship. Conversely, if you were in the forward part of the ship and your GQ station was aft—you ran down the port side. Think of it as driving a car on the right-side of the road—you did not want someone coming at you in your lane. This greatly reduced the possibility of running into other crewmen on their way to their respective stations. Hindering another, or yourself, from getting to one's battle station as rapidly as possible, could mean the difference between life and death. When I and the other members of the gun crew reached our mount, we got into battle dress as soon as possible. The purpose of battle dress was to reduce the possibility of flash burns and more commonly, if you were topside for an extended period, sunburn. Of course, it was not protective armor, but such dress was better than nothing.

Diagram 7-3

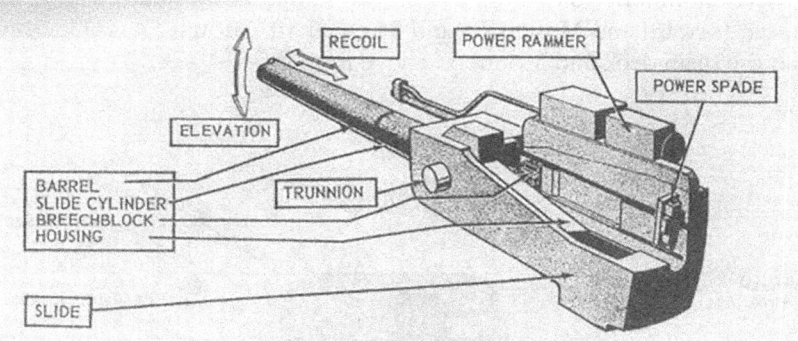

Gun barrel and associated breech housing of a 5-inch/38-caliber mount.

The *Hopewell* employed several different types of gun rounds, including PD (point detonation), "Willie Peter" (white phosphorus), VT Frag (variable time fragmentation), and "star shell" (illumination) rounds. Point detonation was a particularly destructive type of round meant to destroy buildings or sink ships, depending upon your target. Willie Peter was an anti-personnel round that utilized a white phosphorous gel to burn anyone with whom it made contact. These type rounds produced vivid orange centers and fingers of white smoke with orange tips when they exploded in the air, sending burning gel in all directions. They were beautiful to observe if you didn't really think about their incendiary effects. VT rounds were designed to explode in the vicinity of an enemy air target, in order to increase the odds of bringing it down via one of the shell's fragments striking a control surface or puncturing a fuel tank. One or more star shell rounds—a lighted flare suspended from a small parachute during its descent— were used at night to illuminate the battlefield. They could be employed to enable our ground forces to see enemy positions better, or to ensure that a firing ship did not engage friendly forces by mistake during nighttime fire missions.

Off the coast of South Vietnam, our gun crew spent hours on end at our General Quarters stations. At times I found myself, during long periods of inactivity, slumped over the mount's control panel due to heat-induced fatigue. Being constantly hungry and too tired to talk was also common. One summer day, *Hopewell* was given a gunfire mission in support of our ground forces inland. All mounts were firing until one-by-one they began to fail for a variety of reasons. These included an inoperable magazine hoist and a rammer. Mount 51 went down, Mount 52 went down, and finally Mount 54, as well. The only mount still firing was 53, my mount. (Mounts 51 and 52 faced forward and Mounts 53 and 54 faced aft. Mount 54 was located on the main deck and Mount 53 above it on the 01 level.)

Diagram 7-4

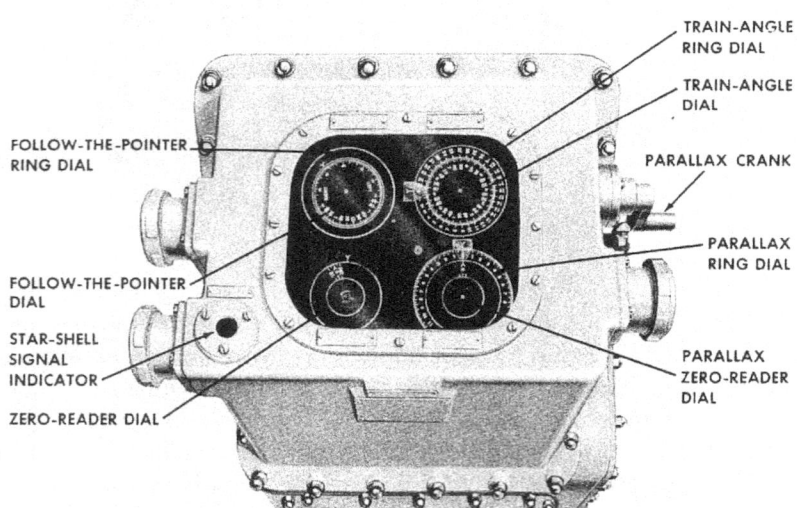

"Crow's feet" indicators used for training the gun.

With only one gun operational, I felt that my job as gun trainer was particularly important. In preparation for firing, I closely monitored the "crow's feet" on the indicator panel in front of me, which told me which way to train the gun mount. This was a more difficult task than one might imagine. My perception of which way I should turn the hand cranks—to move the gun left or right—was opposite of the direction the crow's feet moved. Also, there was urgency to react and to be correct. The indicators were dials housed in a metal box that received electrical signals from the main battery director. These crow's feet signals positioned the mechanical indicators to designate which way the gun was to turn. Such signals reflecting the director's commands were telegraphed seconds before the mount captain gave the order. As trainer, I would then turn two hand-cranked brass wheels either forward or backward to move the gun to port or starboard. It was easy to start turning the wheels in a "knee jerk" reaction, slew the mount in the wrong direction, and have to quickly reverse it.

As I watched the crow's feet moving steadily from the main battery director's preliminary command, I awaited the next command which proved to be "surface action port." I turned the hand-cranked wheels as fast as I could, causing my arms to ache, and when the gun reached surface action port, I threw the lock lever to the "auto" position which gave the main battery director control of the gun mount. In total, we fired nearly 200 rounds at the coast that day.

OTHER ACTIVITIES AND VOYAGE HOME

> *No matter how we try, few of us will be able to forget refueling at sea. On the average of once every two days at sea, we would go alongside an oiler, pass hoses between the ships and take on [tens of] thousands of gallons of fuel. Day and night, while waves crashed along the decks knocking men from their feet, we took on fuel, food and ammunition. In spite of wind and waves,* Hopewell *carried out these operations with an outstanding safety record.*
>
> Hopewell *set two records for refueling on the '64-'65 deployment. First was the fastest rigging time of two minutes. The second record was the fastest rig-unrig time of three minutes and fifty seconds.*
>
> —Description of an under way replenishment from the *Hopewell*'s 1964-65 cruisebook.

When a Navy ship is at sea, it generally refuels every few days. This is not because it has insufficient fuel storage, but due to a requirement to have a high percentage of fuel on board at any given time, in order to be able to respond to a tasking or operation without the need to first find an oiler or enter port to obtain additional fuel. A ship might refuel anytime during the day, daylight was preferable, but it could be in early morning darkness ("zero dark thirty") if necessary. Handling a line in darkness, with spray or sometimes green water that you could not see crashing over you, was not particularly enjoyable. The fear of being thrown to the deck by a wave was always on my mind. An even worse thought was of putting a foot into the bight of a line and losing it, or have the line take you over the side of the ship.

Photo 7-6

Hopewell and the attack carrier *Bon Homme Richard* (CVA-31) refueling from the fleet oiler *Mattaponi* (AO-41) on 2 October 1963.
US Navy Photo No. 1996.488.038.044

Two or more ships steaming in very close proximity to one another was exhilarating to me; though not for the officers responsible for the safety of the ships. One day in the South China Sea, *Hopewell* was refueling from a fleet oiler just after sunrise. We were steaming alongside it, with a hose connecting us providing the life-giving fuel that kept us going. The carrier USS *Kitty Hawk* (CVA-63) was taking fuel on the other side of the oiler. I looked up from the inhaul line I was holding, to witness a very moving sight. Flying in echelon was *Hopewell*'s American flag streaming proudly from her mast; just a little higher was the flag on the oiler; and even higher, the carrier's flag. These flags, flapping tautly in perfect echelon, drove home the fact that we were doing a job that needed to be done in this part of the world.

During this deployment, *Hopewell* became the first destroyer to UnRep (take on fuel or supplies in an under way replenishment) from the *Sacramento* (AOE-1)—a very large ship, and the first of her class of fast combat support ships. This type ship combined the functions of a fleet oiler (AO), ammunition ship (AE), and a refrigerated stores ship (AFS). She was known as a "floating supermarket" because of all the goods she carried in her AFS role, and was also quite fast, having old battleship engines for propulsion.

THE TRADITIONAL CRUISEBOOK

Image 7-7

Mt. 33, Sky 4, Commence Firing

The term is NEGATIVE, not THE HELL I WILL!

It is customary for a handful of crewmen of a ship on deployment to create a cruise book under the oversight of a junior officer. These hardbound, treasured keepsakes normally include group and individual photographs of everyone aboard, as well as photos of ship operations and liberty ports. Sailors enjoy humor, particularly during the endlessly long days of a deployment. Thus, if the ship has a cartoonist among its officers or enlisted ranks (almost always the latter) some of their work will normally find its way into this tome. The cartoons (above) are representative of those I drew for *Hopewell*'s 1964-65 cruise book, which typically represented the activities of different divisions. The one on the left is devoted to Fox division, which was made up of the fire control technicians who controlled the guns. The officer in charge of the secondary gunfire director (SKY4), is calling for the dual 3-inch/50 mount (Mount 33) to fire and is obviously not aware the gun turret is going to blow off their heads, but the terrified crewman is. In reality, the mount had redundant electrical and mechanical "stops" to prevent the gun barrels from slewing toward, and firing into, any part of the ship. The second cartoon is for the Operations/Communications division. One of the guys in the radio shack is correcting the other regarding the proper use of naval terminology.

HOMEWARD BOUND

How much better the land seems from the sea than the sea from the land...

—A Spanish official, after crossing the Atlantic, a much smaller ocean than the Pacific, in 1573.

At the completion of combat duty, we were released from Seventh Fleet and began the long Pacific crossing to San Diego. Our last stop on the voyage home was at Pearl Harbor for fuel, provisions and crew liberty. On a shore run, I and some of my shipmates made our way to a Waikiki beach and purchased a case of beer, some of which we consumed before piling into a rental car. We then embarked on a cross-island road trip to visit an acquaintance of one of our group. It had been raining and the narrow road we were on would have been hard to navigate, even without being plied with drink. Sliding around corners and powering up straightaways was neither smart nor particularly beneficial to our already queasy stomachs, but we all survived. In different circumstances, we would have enjoyed the spectacular sights along the way, which included waterfalls, lush vegetation and deep ravines—in which we could have very easily found ourselves. Compounding our poor judgment was showing up unannounced to visit a lady friend of our shipmate, with whom he had once worked. As she served us tall screwdrivers with lots of vodka and just a hint of pineapple juice, I wondered what her poor husband thought about four drunken sailors invading their happy home that evening. He must have been a hell-of-a-guy. Our good fortune continued as we, somehow made it back down the long and winding road to Pearl Harbor.

Nearing the West Coast on the final leg of our trek home, we picked up radio station KEWB (which is now KFRC) from the San Francisco Bay Area playing rock & roll. The first song we heard was "Wooly Bully" by Sam the Sham and the Pharaohs. Not a big favorite. Comments made by some of the crew included, "If that's the kind of music being played in the States, let's turn around and go back!" Later we heard Dionne Warwick's "Message to Michael," which was great and the music being played on the airwaves just got better and better.

Hopewell entered San Diego Harbor on 6 February 1965. I was very excited as she proceeded inbound to the naval station. I had leave coming, and was expecting my parents to meet me on the pier and to drive me back to Chico. Eager to see me, my mother exclaimed to my dad, "There's Michael's ship," as each ship in our flotilla approached the naval station. The first ship she saw was the cruiser *Columbus* (CG-12), known as the "The Tall Lady." She was the flag ship of our cruiser-destroyer flotilla (FlotEleven) and was the first in line of our eleven ships. My mother thought each succeeding ship was *Hopewell* until after about the fifth or sixth to come into view. As luck would have it, our commanding officer was the junior in rank and therefore *Hopewell* was the last ship to moor alongside the pier. During the 186-day cruise, we had burned 2,236,423 gallons of fuel oil while steaming 42,841 nautical miles. To keep the ship looking pristine, the boatswain's mates and seamen had used 760 gallons of paint, and they and other members of the hard-working crew had downed 26,793 sodas. Of course, these numbers were relatively unimportant to us; most importantly we had enjoyed a total of sixty-three days in liberty and/or working ports.

8

A Long-Awaited Visit Home

Mike I have my learner's permit, I have been driving for a little while, and I am pretty good. Will you take me out to the airport to practice my driving?

—Innocent request by my brother leading to the wrecking of my car.

Following the return of *Hopewell* from deployment, I spent some well-deserved leave in Chico. Since my hometown was a long 600 miles north, I usually spent my weekend liberties in Long Beach, 100 miles north of San Diego. "It was the winter of '65; I was hungry and just barely alive," I can still hear those lyrics by Joan Baez in "The Night They Drove Old Dixie Down." It was actually February 1965, and not 1865, and I was home; time to get caught up with old friends, try to meet girls and do some serious drinkin'.

Before enacting any of these plans, my brother Alan asked me if he could use my car to practice his driving. My first mistake was to assume that since my car had been at my parents' house while I was on deployment, he had been driving it in my absence. I found out the hard way that he had been learning to drive using my parents' car. This mattered, because while my parents' car was a '57 Pontiac with an automatic, mine was a '56 Chevy with a stick.

Alan got into the driver's seat and I took the passenger's side. We were in my parents' driveway. He turned on the ignition and killed the engine. With some coaching, we backed out of the driveway and shot straight across the street to a neighbor's driveway where he fortunately killed the engine before we crashed into their house. It's a good thing no one was coming down the street as we crossed it. After he restarted the engine, I told him to ease off the clutch as he gently pressed down on the gas. Great advice, but he accidentally popped the clutch instead, which scared him and he stomped on the accelerator, not the brake.

Now we were hurtling, at hypersonic speed, straight for my parents' car. I grabbed the steering wheel, and shoved it hard left.

Instead of plowing straight into the Pontiac, we glanced off it doing hundreds of dollars' worth of damage to the rear quarter panel instead of totaling the car. My poor ol' Chevy continued its path in a semi-circle until it reached the curb on the same side of the street as our neighbor's driveway. We jumped the curb, and wrapped the front end of the car around a tree in another neighbor's yard two houses down. My head was thrown forward and then backward leaving a Brylcreem (hair gel) spot from my head on the windshield. Alan was shaken up, and a little dazed. As he sat trembling behind the wheel, I got out to assess the damage. To the assembled multitude of neighbors that had flocked around to see what had happened, I remarked "And for our next stunt!" upon which Alan's lip began to quiver. Looking across the street at my parents' car, I then remarked, "Well at least we kept it in the family!" That was all it took for my poor little brother to bolt from the car and run home crying. In retrospect, I should have driven out to the airport and utilized its wide roads with only sparse traffic to assess my brother's driving skills. Following the accident, I should also have respected his feelings more and not worked the crowd with my cornball stand-up routine. Alan eventually forgave me for my lapse in judgment and insensitivity.

Speaking of "on leave" disasters, I was now in the market for a date or two. My girlfriend of three years had sent me a "Dear John" letter a few months before, and thus I was available for any and all girls in Chico, except, obviously, her. I made a few phone calls to girls with whom I had graduated from high school. Most were married or engaged. I also called a few girls from my sister's class, and what a surprise, I apparently was damaged goods due to the unpopularity of the war. There was an anti-war poster at this time that exclaimed "Girls say yes to guys that say no," which I truly hated.

While home on leave, my dad and I watched the TV series "Combat," starring Vic Morrow and Rick Jason. The plot of one particular episode involved a squad of GIs in hand-to-hand combat with German soldiers. I turned to my dad and remarked that I could not engage in close quarters combat, and that I was glad to be serving in the Navy aboard a ship. He responded, "Mike, if you had to, you could, and you would." Many years later I learned about some of his combat experiences from my brother-in-law Bruce MacDonell. He and my dad would occasionally stay up until the wee hours of the morning drinking beer and talking about anything and everything. The beer apparently loosened Dad's tongue, and he told Bruce things that he had never shared with me. I thank Bruce for not only giving my dad a time to come to peace with his war experience, but also for relaying to me what my father had gone through.

On one of my visits to Chico, my parents invited me to go with them to a VFW (Veterans of Foreign Wars) dinner with featured speakers. Once there, Dad introduced me to members and guests as being in the Navy and just back from Vietnam. The assembled multitude accepted me and made me feel welcome. My dad was a member of VFW Post 1555 and I was quite proud of him. He had been a POW during WWII after being captured in the Battle of the Bulge. He was twice wounded in the battle and almost died of malnutrition in Stalag IX-B (in Bad Orb, Germany), reportedly one of the worst camps regarding its treatment of prisoners. I admired him for many reasons, including not letting the bad things that happened to him effect his life in a detrimental manner, and especially for being a very compassionate and positive person. Later, that night, the post commander honored him with a lifetime membership to the VFW. After expressing his appreciation, my father gave a twenty minute extemporaneous speech on what it meant to be an American and a patriot. My eyes welled up and I choked back a few tears as he spoke. Everyone in attendance was greatly moved, and I could not have been more proud. I had not been unaware that my soft spoken father was a gifted public speaker; his performance as an orator was quite a surprise.

Bruce told me that Dad once found himself face-to-face with a German soldier. Each had his weapon trained on the other and it was kill or be killed. Dad fired first. I can't imagine how he must have felt, and I hope that I never find out. This revelation also made me realize that my father could easily have failed to return home from the war. Since he did, his experiences might have made him bitter against all Germans. To his credit, he did not hold a grudge. A German family moved into our neighborhood, just two doors down, and my father welcomed them as he would have any others. We were to find out, later, that they escaped from East Germany and had suffered much. Their son was in my class at Chico High. The father had been in the Wehrmacht (German Army), just as my father had been in the U.S. Army. Perhaps he had at one time believed in Hitler's Nazi philosophy, or more likely he was forced into the army, as many were.

74   Chapter 8

Photo 8-1

Barracks at Stalag IX-B, Bad Orb, Germany.
(Courtesy of www.LoneSentry)

One of the highlights of my visit to Chico was going out for a beer in my dress blues with a friend at a new restaurant, the Italian Cottage. The owners, Burt and Judy Katz, furnished me a gratis pitcher of beer. I have never forgotten that act of kindness they bestowed upon me, especially with the way that the general public treated us. While we were enjoying the beer, they informed me that because of the lack of customers they would probably have to go out of business. Given their financial situation, it must have been hard for them to give away that pitcher of beer. But they did it anyway. A few years later, I went by that location to see what had happened to the Italian Cottage, fully expecting to see another business in its place. To my great surprise the adjacent Signal gas station had been removed to make room for a parking lot to handle all the customers at a now thriving Italian Cottage. I guess karma is alive and well in the universe. The Italian Cottage is known for having the best pastrami sandwich north of San Francisco, and I will never forget the kindness extended to this serviceman.

Eventually it was time to leave Chico and resume Navy life. As I was standing in line at the Greyhound Bus station awaiting transportation to San Diego, a little old lady, upon sighting my dress blues, came up to me and asked if I knew her grandson, since he was in the Navy, also. She told me his name, and was disappointed that I did not know him, apparently not realizing what a huge organization the Navy was, particularly during wartime.

9

Liberty in San Diego and Long Beach

I have been in worse places than this.

—Kit Carson, speaking about San Diego

At this point, readers have probably deduced that I, like many sailors, liked spending time in clubs and bars drinking. Unfortunately, such pleasures require disposable income. One might suppose that since sailors don't usually frequent expensive establishments that I should not have needed additional money. However, junior enlisted men did not make very much money, and many visits to cheap bars drain one's wallet just like a few better establishments. Wanting extra money to enjoy myself, I started a "slush fund" aboard ship. Anyone who has ever served in the Navy is familiar with slush funds, the proprietor of which is a sort of loan shark. I was involved in this activity aboard the *Hopewell*. A shipmate would come to me and borrow $5.00, then pay it back plus interest, two weeks later. He would pay me $7.00. If he borrowed $10.00, then he would pay $14.00; the return on my money was forty percent. The same people were always without money when liberty came around and borrowed continually. For some reason, it did not occur to them to simply stay aboard for one or two weekends and thereby eliminate their running debt.

Fairly quickly, I realized that I could not enforce collection if someone did not want to pay me. The reason for this was I was nowhere near being one of the biggest or meanest guys aboard the ship. As a yeoman, I was on many occasions in charge of issuing liberty cards and/or passes, which along with a military ID, allowed an enlisted man to leave the ship on liberty. In theory, I could keep my clients from liberty if they did not pay me; in practice that didn't always work.

One day, the Chief Master-At-Arms, which made him the head "lawman" on board, came up to me and said, "Hey Halldorson, I hear you have a slush fund." This sent shivers down my spine; since it was a violation of the UCMJ (Uniform Code of Military Justice) to run a slush fund, and if caught I would face "Captain's Mast" and likely be awarded severe punishment. I answered truthfully and said, "No, Chief, I do not; I shut it down last week." This was due to my fear of not being able to collect monies due. The chief looked perplexed and then said with a hint of disgust and slight resignation in his voice, "Damn, I needed some money for liberty!"

LIBERTY IN LONG BEACH

In summer 1965, the *Hopewell* was in the Long Beach Naval Shipyard from 15 August through 15 September for some much-needed work. When not aboard ship, I greatly enjoyed running the streets of the city with Jim Brickey. We were good friends, were both yeomen, and were both assigned to the same gun mount for battle stations. Jim's parents lived in Long Beach and he and I had previously routinely visited the city when the *Hopewell* was in San Diego. We would travel north from San Diego on weekends in his big blue 1958 Oldsmobile convertible. After a weekend liberty, Jim and I would get up early Monday morning around 3 a.m. to return to the ship and be aboard the *Hopewell* by reveille (6 a.m., the time at which the crew was awakened each morning). The first time I rode with him we agreed that he would drive half way to San Diego and I would drive the other half.

I put my head down to sleep for the first half of the hundred-plus mile commute. For some reason I awoke from my slumber, looked over at Jim, and was shocked to see him slumped over the steering wheel. The car was doing 80 miles an hour, and I yelled at him to wake up. He did so, and after he had put his head out the window and slapped himself in the face several times, I tried to sleep again. Still ill-at-ease, I glanced up at him a few minutes later and he was again slumped over the wheel. Jim never drove us back to the base again; I always did so. He later told me that, before I began to commute with him, he had once been awakened by the sounds of his car scraping the car next to it, at highway speeds, while returning from liberty. After his stint in the Navy, Jim went to Long Beach City College and then Humboldt State College, where he obtained a degree and became a teacher. He would become the principal of a school in an Alaskan village and, after eventually returning to the "lower 48," became the mayor of his home town in South Dakota.

I spent a lot of liberty time in Long Beach, which was then a big Navy town, akin in many ways to San Diego. I would prowl the city streets and adjacent southern California areas in search of girls. One of my journeys took me to Greg's South Street A-Go-Go in

Bellflower, or perhaps in Lakewood, I don't recall. I was looking forward to seeing a new rock & roll group, but had too much to drink and was awakened under a table at closing time. I can still visualize the banner which hung across the ceiling of the stage announcing Sonny and Cher's new song, "I Got You Babe!" Other people found other ways to have fun. One day, while walking down a sidewalk, I sighted a businessman in a three-piece suit and a martini glass in his hand. He was gliding along on what I was to find out was a skateboard. Jan and Dean had a hit song called "Sidewalk Surfing" which had come out in 1964, but I never really knew what it meant until then.

My principal companions in the Long Beach-Los Angeles area were my shipmate Jim Brickey and my cousin Patty Stanfield who lived in Bellflower, near Long Beach. Patty had a 1955 two-toned red and white four-door Buick, while Jim's "big blue tank" was a 1958 Oldsmobile convertible. For obvious reasons, we were almost always in the convertible. Friday and/or Saturday nights would find us cruising the streets in Los Angeles or Hollywood. We would drive by the "Whisky A-Go-Go," or other such places, just to see "the scene." Coming out of the car radio would be Beach Boys' songs like "Fun, Fun, Fun," "I Get Around," "California Girls," and a new song "Red Roses for a Blue Lady" by Wayne Newton—an artist unknown to us. At first we thought the latter song was sung by a girl. Newton was then very young, and his voice had not yet changed.

Photo 9-1

Two-tone 1955 Buick, like the one my cousin Patty owned.

Patty had a girlfriend, Janie, to whom she eventually introduced us, and the four of us had a ball just cruisin' around taking in the sights. No one drank on these nights, because the girls did not imbibe. Invariably, we would end up at a Denny's and the standard order was a patty-melt and coffee. I cannot remember any of the subjects of our conversations, but I do remember the constant laughter. The laughter was a welcome release from the toil associated with shipboard life and sea duty.

Patty's parents were very kind to me, allowing me to stay at their home and use it as a base of operations. They would treat me to Kentucky Fried Chicken and let me "veg" in front of their TV. There was a show I liked, called "Hullabaloo," which featured the surfing music of the day and scenes of surfing. On nights we were not with the "the girls," Jim and I would usually be in a night club, dancing to songs like "The Land of a Thousand Dances" by the rock band Cannibal & The Headhunters.

Photo 9-2

The place to be – Furs, hats and straight black ties – a view driving by night club "Whisky a' Go-Go" in the 1960s.

LIBERTY IN SAN DIEGO

In Navy parlance, I was a "steamer" (someone who liked to hang out in bars), so I also spent much time enjoying the nightlife in San Diego. (Its downtown area in the '60s was very different from today. San Diego is now known as "America's finest city." Not so, when I was in the Navy.) The downtown area was populated with bars and locker clubs. The locals considered these establishments seedy and generally preferred that they—and the sailors and Marines who frequented them—not be there. Locker clubs were places where sailors, who were required to be wearing uniforms when departing and returning to their ships, could store their "civvies" (civilian clothing). Locker clubs also had laundry facilities and other amenities. You could get alterations to your clothing, and condoms were available in vending machines—everything you needed to make for a successful liberty or leave.

Photo 9-3

The Seven Seas locker club in downtown San Diego near the waterfront.

As much as we tried to pass as civilians, there was no getting away from those shipboard haircuts. Most girls could tell us a mile away. In Hawaii, some locals used to chant as sailors on liberty walked by, "tides out, seaweed in, anchor-clanker, swabbie." Probably, the area most renowned among Navy men for hostility toward sailors was Norfolk, Virginia. Common signs in yards, asserted "sailors and dogs keep off the grass." In retrospect, given the antics of many young sailors where huge populations of servicemen could overwhelm the local area, I can understand their sentiment. At the time, however, it made me angry.

One Sunday morning a couple of us found a bar open in Chula Vista, a low-income city in south San Diego frequented by sailors. We decided to try it even if it was a western bar, a type of watering hole we would not normally frequent because we liked rock & roll music. Conversely, they would have alcohol. Inside there was an all-girl Western band on stage. The girls were smilin' and tasseled-up with frilly cowgirl sleeves, hats, and skirts; workin' the crowd, asking for requests for favorite songs. Who knew any western songs? We certainly did not. The lead singer issued a challenge to the audience. Anyone who could stump the band with a request that they could not perform would get a free drink. However, if they knew it, and played it, the requester had to buy everyone in the band a drink.

They presumably expected customers to request cowboy songs. Thinking I could stymie the band, I yelled out "Satisfaction," a song by a relatively new British group, the Rolling Stones. The leader flashed a knowing and slightly sarcastic smile at me and the band started to play. They started the song alright, but could they get every word and every note right? They performed it with perfect lyrics and I was out seven drinks, a chunk of my minuscule pay check, and a considerable piece of my pride.

As good as liberty in San Diego was, homesickness would periodically send me north to Chico. On one such occasion, I set off for home on a four-day liberty, but took a bus, instead of flying, to save money. However, due to heavy rains and the possibility of mudslides, the bus route was altered to "the long way around." Instead of going straight up Interstate 5, we headed east into desolate and rural desert areas of southern California, and made many unexpected stops. One was an hour layover at San Bernardino, my birthplace, which I had left as a newborn baby. Perfect, I could visit the town and tell my parents what it looked like now.

As I exited a store onto the sidewalk, I was greeted by a loud engine roar and was almost run over by a motorcycle—on the sidewalk. And to make matters worse, the guy on the chopper was wearing a gang jacket with "San Berdoo" on the back and had a menacing smile. What the hell kind of place was this? After I collected my wits, a feeling of gratitude came over me that my family lived in Chico; a friendly, picturesque, and relatively innocent rural town.

SHORE PATROL DUTY

The opposite of sailors drinking and "cutting up" was performing Shore Patrol duty. The Navy, being very wise regarding the habits of young men, particularly if they had been at sea for some period, took measures to keep them in line when on "runs ashore." The Army and Marines had MPs (military police), we had Shore Patrol. The latter title referred to members of a ship's in-port duty section, chosen to function as overseers of good behavior by shipmates who, not having 24-hour duty that day, were free to go ashore after the end of the work day.

I was once assigned to the Shore Patrol at a *Hopewell* ship's party. I had an official dark blue band with the letters "SP" in vivid yellow affixed over my upper arm, and was armed with a nightstick to break up fights, if necessary. I felt pretty important and was likely strutting like a peacock. As I walked about maintaining good order and discipline, I came upon our operations officer, Lt. George Cooper. He held out a bottle of beer and said, "Halldorson tell me if this is alcoholic." I took a sip, despite being prohibited to do so, and pronounced it alcohol. Lieutenant Cooper then said, "No, drink the whole bottle and let me know for sure." My doing so confirmed my diagnosis. After that, I kind of floated around, and noticed that the chief petty officer who was in charge of the Shore Patrol was passed out in the corner with a bottle of booze in his lap. As my rounds returned me to the table at which Lt. Cooper sat, he once again ordered me to sample some beer, which I did and I became even loopier. I don't remember too much after that, but had there been trouble, it is unlikely that any of our shore patrolmen, who by then were in a similar condition, would have been able to deal with it.

The fact that our officers allowed, even encouraged, us to drink attested to their belief that there would only be goodwill among *Hopewell* sailors that night. Sailors normally avoided senior officers, but there was some comradeship between junior officers and the crew. Once, while we were in port, the *Hopewell* put on an intramural softball game. Usually ships played each other, but not on this occasion. The teams were the officers against the Operations Department. We won a squeaker with a score of 23 to the officers' 22. I had played second base against my will; I was an outfielder and would be all my life. But, anything for the team, particularly if it meant beating the officers.

ADVANCEMENT TO PETTY OFFICER

While enjoying time ashore, I was also working very hard aboard ship and thus earned advancement to petty officer third class. Navy non-rates yearn to advance to petty officer to gain additional standing within their commands and to, hopefully, avoid menial tasks typically levied on junior enlisted. Advancement is based on the quality of your day-to-day work, and how you do on a Navy-wide test. To advance in rank in the Navy, you must compete with your peers across the entire Service, unlike in the Army in which local base commanders promote soldiers. After studying for and taking the test, you must wait to learn if your combined score is high enough to secure one of the available spots.

The day finally came for me to "tack on my Crow," Navy slang for sewing on a rating badge depicting an eagle, with one or more chevrons below it. Since I was a yeoman, there were two-crossed-quills immediately below the eagle signifying my rate. Now I wouldn't have to put up with as much of the abuse that yeomen commonly received from shipmates for doing clerical (girl's) work. Funny we did not get such when they wanted their liberty cards. Being an artist, I not only sewed my new crow on my sleeve, I drew "crows" on my "skivvies," Navy issue underwear, as well.

10

Second Deployment

If we are driven from the field in Vietnam, then no nation can ever again have the same confidence in American promises or American protection.

—Statement by President Lyndon B. Johnson at his White House press conference of 28 July 1965.

In early 1966, the *Hopewell* was preparing for another deployment to the Western Pacific for combat duty off Vietnam. The requisite training included participation in shore bombardment exercises off southern California. After enjoying relatively balmy weather, we then travelled north for Operation BUTTONHOOK, a joint American and Canadian exercise off the west coast of our two countries which emphasized antisubmarine warfare techniques.

Prior to this deployment there was increased trepidation among *Hopewell* sailors about going to Vietnam. Although we had previously served off the coast of the Republic of Vietnam, the war effort had dramatically increased in scope since then, with an associated increase in American casualties. Thus, I was understandably concerned about sailing off to war, and the possibility of death, which spurred greater interest by me in organized religion. I had only gone to church as a child, then as a youth, and that was only because my mother made me, but I accepted an invitation from my cousin Patty to go with her to her church and to "get on the right side of the Lord."

Patty was a member of the Four Square Church, which to say the least, was different from the Trinity Methodist Church, in Chico. The congregation was nice and greeted me warmly. I was in my dress blue uniform which had my ship's name patch on the right shoulder. One of the men in attendance, who was seated directly behind me, tapped me on the shoulder and told me that he had served aboard the *Hopewell* in the mid-fifties. What a coincidence. Was this a sign?

When the service started, I was surprised to find they were talking in tongues and shouting out loud. Also, when the spirit moved them, individual people stood up independently of others in the service, unlike the practice at the Methodist Church. My attendance was, however, a small price to pay if it helped me survive the war.

The *Hopewell* stood out of San Diego Harbor on 1 March for duty with the Seventh Fleet. En route were stops at Pearl Harbor, Midway Atoll, Yokosuka, and Subic Bay for food, fuel, and repairs before arrival off the coast of South Vietnam. During this deployment, we spent approximately three-and-a-half months on the gunline, delivering firepower in support of ground forces, firing 2,276 rounds at 103 targets between 31 March and mid-July 1966. Some of the targets were enemy facilities and others, troop ground positions. In addition to destroying 112 structures, *Hopewell*'s shore bombardment damaged 550 bunkers, silenced enemy mortar and automatic weapons fire directed at friendly forces, and routed three battalions of Viet Cong.

Many aspects of my second deployment to Vietnam were similar to the first, but there were some significant differences in ship operations, and for me personally, increased maturity. The same could not be said for our crew, generally, and those of the other ships in our division. In liberty stops across the Pacific, members of our crew fought with sailors from other ships comprising Destroyer Division 172: the *Lynde McCormick* (DDG-8), *Parsons* (DD-949), *Porterfield* (DD-682), and *Agerholm* (DD-826). A huge brawl took place in Pearl Harbor, the first stop on our trek across the vast Pacific. This could have been due to a variety of factors, but it was probably caused by a change in crew composition between deployments. Following a ship's return from a cruise, many crewmembers would often leave the Navy, their enlistments being up, or be transferred to a new command.

PACIFIC CROSSING

After several days at sea, our next stop was at Midway Atoll, the western sentry of the Hawaiian Islands, for refueling. During this brief four hours in port, most of the sailors went ashore and drank themselves silly. At my and other tables in the EM Club, stacked empty 16oz beer cans reached toward the ceiling. Sensing impending trouble after enough sailors got "well-oiled," I left early and returned aboard ship. Along the way, I took pictures of gooney birds taking off and landing.

These type fowl run and run to get up enough speed to finally lift off and become airborne. When they land, they flap their wings as if using air brakes, violently plant their feet, and usually go "ass over tea kettle." I got close enough to a pair of baby gooney birds to snap their picture. As I pushed my camera into their little faces one of the birds, obviously not awed by a strapping American fighting man, bit my hand.

After reporting back aboard, I stood on the fo'c'sle with one of the officers and watched him fish. We were having a friendly conversation when the claxons sounded for the crews to return to their respective ships. As a wave of sailors swayed down the dock, a huge fight broke out. Several guys went off the pier in their undress blues. The turquoise water was so pristine that I could see the underwater hull of the ship alongside us. This same water was now cluttered with thrashing bodies. One can only wonder about how these idiots felt—after too much drink, being struck by fist or foot, and being dunked—being told to re-man their sea and anchor detail stations to get under way. Many would then have to take up regular under way watches.

After leaving Midway Atoll, we stopped at Yokosuka, Japan, and Subic Bay in the Philippine Islands. The *Hopewell* then pointed her bow west, bound for Vietnam.

Photo 10-1

Gary Lyle and I enjoying several "cold ones" at the EM club in Subic Bay, The Philippines. Don't we just look happy as hell? Wasted days and wasted nights!

ON THE GUNLINE

> *As the sun disappeared below the horizon and its glare no longer reflected off a glassy sea, I thought of how beautiful the sunsets always were in the Pacific. They were even more beautiful than over Mobile Bay. Suddenly a thought hit me like a thunderbolt. Would I live to see the sunset tomorrow?*
>
> —Eugene B. Sledge, *With the Old Breed: At Peleliu and Okinawa.*

It often seemed incongruent to me that while in the combat zone one might enjoy a beautiful sunset while at the same time worry about very real danger that had nothing to do with nature. While in our patrol sector off the coast of Vietnam, we would listen to "Hanoi Hanna" at night. The propagandist gleefully informed us that students back home were burning their draft cards, and wondered aloud what we were doing steaming around out in the Pacific instead of in a warm bed with a girl. After the war, it was learned that the radio personality was a Vietnamese named Trinh Thi Ngo (born in 1931), but, at the time, no one knew her true identity. Nevertheless, we did not appreciate the North Vietnam-sponsored English-language broadcasts directed at U.S. troops.

The *Hopewell* carried out many Naval Gunfire Support (NGFS) missions off the coast of Vietnam. Some missions were pre-planned; others were in response to ashore "calls for fire" to drive the enemy back from friendly forces. The shore bombardment score for a particular mission, or missions, tallied how many structures we destroyed, how many we damaged, sampans destroyed, bunkers destroyed, and enemy killed. Reviewing the scores was a little surreal. They were not Little League scores or report cards of your school grades, but instead a metric of death and destruction summarizing the effectiveness of our gunfire. The *Hopewell*'s shore bombardment totals on 15 April 1966, after fourteen days on station, and 1,556 destructive rounds, were as follows:

Target		Target	
Structures destroyed	73	Bunkers destroyed	4
Structures damaged	366	Sampans destroyed	6
Structures left burning	465	Personnel casualties	1
Trenches destroyed	2	Secondary explosions	2

In addition, we fired one hundred twenty-two illumination rounds during eighteen night missions, to pierce the veil of darkness.

Looking through the crosshairs of my magnified gun sight in Mount 53, I was able to see our rounds hit, unless the entry point into the jungle was far upslope from the beach. Most of our missions were in the daytime, and the weather was usually sunny, hot and muggy. The beach in my crosshairs could have been a Hawaiian island, had it not been viewed through a round site with perpendicular lines interrupting the view.

Photo 10-2

Two 5-inch/38 gun mounts like the four aboard the *Hopewell*.

One day, I watched a Viet Cong (a member of the North Vietnamese Army) running with a rifle along a shallow beach at the tree line. He came into view on my gun sight from the right and as I watched him proceed across my scope, a round from my gun mount hit the beach, right where the figure had been, seconds before. There was an explosion with sand and trees flying through the air. Where the man had been before, he was no more. Although I did not identify him as a target or fire the mount (a responsibility of the pointer), that memory has not left me. Visual flashbacks occur when least expected, sometimes interrupting my sleep at night.

The explosions of white phosphorus incendiary rounds were beautiful to look at from a distance, but not if you were the target.

Another of the *Hopewell*'s naval guns (mount 51) was firing at a group of men (believed to be the enemy) launching sampans. The Viet Cong used these type of indigenous craft to offload weapons and munitions from ships (North Vietnamese gunrunners), and transport these arms inland by water. Five rounds of white phosphorus and five more of regular destructive fire dispersed the group; survivors ran back into the jungle. The sampans remained on the beach, shot up, their mission thwarted that day.

The military newspaper *Stars and Stripes* later described the operation of which the *Hopewell* had been a part, under the headline "Vietnam Diary," excerpts of which follow:

> April 5th – Marines continue Operation JACKSTAY south of Saigon, but reported no significant contact. Total Viet Cong killed in the 10 days of the operation now stands at 62....
>
> The destroyer *Hopewell* fired 201 five-inch rounds at VC structures and troop concentrations in the I Corps area.
>
> April 12th – In the I Corps area, the destroyer *Hopewell* fired 324 five-inch rounds against VC targets. Air spotter reported the fire was quite effective. In addition, gunfire from *Hopewell* silenced a VC mortar and 2 recoilless rifle positions and assisted in routing a large number of VC troops.
>
> April 15th – One VC was confirmed killed as a result of fire from the destroyer *Hopewell*. The *Hopewell* fired 49 five-inch rounds at VC positions in the I Corps area with outstanding target coverage.
>
> April 16th – The destroyer *Hopewell* fired 124 five-inch rounds at VC targets in I Corps. Airborne spotter reported 10 structures destroyed, 14 damaged and 4 left in flames.

OPERATION LAMSON 255

During April 1966, the *Hopewell* participated in Operation LAMSON 255, which included providing naval gunfire support for the ARVN 1st Division in Quang Tin Province. This South Vietnamese unit, the 1st Division of the Army of the Republic of Vietnam (ARVN), was assigned to I Corps, an area encompassing the northernmost region of South Vietnam. The 1st Division was based at the old imperial city of Hue, the Corps headquarters. During this operation, the *Hopewell* was, for some forty-eight hours, positioned only twenty miles south of the 17th Parallel, the line separating North and South Vietnam.

The following month, the crew of the *Hopewell* learned on 23 May that we had been assigned as "gun support ship" for the light cruiser *Oklahoma City* (CLG-5). The cruiser was serving as "Tom Cat," a reference point for carrier aircraft returning from strikes over North Vietnam. During this assignment, we were in "Indian Country," positioned seventy-five miles south-southeast of Haiphong, North Vietnam, and eighty-to-ninety miles west of the island of Hainan, a part of Communist China. A note in the Plan of the Day enjoined the *Hopewell*'s crew to be vigilant:

> HOPEWELL is within easy range of NVN [North Vietnam] and CHICOM [Communist Chinese] aircraft, PT boats and other craft. Expect to make sightings of communist fishing craft in the area. Because of the various threats facing us, it is imperative that lookouts, signalmen, quartermasters, gun crews, radar and ECM [Electronic Countermeasures equipment] operators maintain the highest level of alertness in order to gain early detection and prosecution of any contacts which may appear. HOPEWELL may be called upon to provide gun cover for and participate in SAR [search and rescue] missions. As many of you know, some SAR support ships have come under fire from shore batteries when covering recent missions, a particular reason for alertness.

LIBERTY IN JAPAN

Sailors deployed to the Western Pacific usually refer to such duty as being in "WestPac" or on a "WestPac cruise." For Vietnam era sailors, our Western Pacific cruises, once in the Far East, were split between duty in the combat zone and liberty/maintenance periods in the Far East or at Subic Bay, Philippine Islands. Liberty ports were our form of R&R (rest and relaxation).

On one such occasion, a couple of us were in a bar while on liberty in Yokosuka and were looking good in dress blues with our liberty cuffs. I had gotten dragons embroidered inside the cuffs of the sleeves of my jumper. No harm, no foul to the U.S. Navy if hidden from view. Fortunately, although not sanctioned by the sea service, displaying such was ignored, as long as it did not happen aboard ship or at official functions. My cuffs were rolled up for everyone to see: red, green and orange dragons.

I walked over to the jukebox, inserted a quarter and played my favorite song, "One, Two, Three" by Len Barry. After it ended, I was making another selection when a bar girl came over and stood next to me. The bar was very dark, the only light source being the neon glow from the jukebox. She looked at me and said, "Sailor boy, you would be good looking if it weren't for the bumps on your lip." That was a mood breaker. I had a skin condition that produced small bumps under the skin on the upper lip. That girl, if she was trying to get me to buy her drinks, went about it all wrong. She needed to work on polishing her customer skills, but I still enjoyed my night on the beach.

The following morning, after a single day and night of R&R at Yokosuka, we put to sea bound for Sasebo. This port city on the island of Kyushu, Japan, hosted a U.S. naval repair base acquired at the end of World War II. After arriving there, I happened to look across the harbor and sighted the *Zelima* (AF-49), a refrigerated stores ship better known as the *Zippin' Zee*. Keith Girard, a friend of mine from Paradise, California (located on a ridge sixteen miles from Chico), was assigned to her. I called over to his ship, and he agreed to meet on the base at the EM club for a steak dinner, then go into town to visit a couple of bars. After finishing our dinner, we took a cab to Sasebo City. Upon spying the Paramount Club, the first watering hole we came to, we promptly ordered the cabbie to stop and let us out.

The bar was dimly lit with an overhead spotlight producing a single shaft of light near the center of the room. We found a somewhat illuminated table, got our beers, and were treated to a young lady who danced on our table. After drinking for a while, I needed to use the bathroom. As I opened the door and stepped into the bathroom, it closed behind me and I heard, "Your ship sucks!" Then the retort, "No, your ship sucks!"

A shipmate, Garvey, was at the far urinal shouting at three guys to his right. I said, "Garvey, shut up. Let's get out of here," feeling it my duty as a petty officer to protect a lowly seaman apprentice. I also felt compelled to show him how a *Hopewell* sailor behaved in public. Unfortunately, the damage was already done. As I uttered the words "Garvey, shut up," one of the guys offended by Garvey's insults threw a punch at my nose and connected with it. When I woke up, all I could see through tears generated by the blow was a section of the floor covered with blood—all of it from my nose. Garvey lay on top of me. He was not bleeding, but actually got the worst of it having been pounded pretty good.

Keith, waiting for me to return from the head, watched the Shore Patrol carry a sailor out by his arms and legs. He shortly realized the sailor was me. I was taken to the base hospital and only later learned what else happened that night. It seems that after witnessing what happened to Garvey and me, others from the *Hopewell* followed the offenders all over Sasebo where other fights occurred.

The doctor and his staff at the hospital, already "lightly loaded," were watching a movie, which was stopped while I got my damaged nose tended to. The doctor pushed it back in place, taped me up and returned to his movie, a western. In the scene being shown, the actors were in the middle of a huge barroom brawl with fists flying. Faces were being pounded, chairs were being broken over heads, and bodies were flying all over the place. At no time was there any blood visible on any of the participants. I took one punch and lost what I felt was a gallon of blood, but it was likely only a few ounces.

The next day I returned to the base hospital. The doctor readjusted my nose, and gave me morphine to ease the pain. Thereafter the orderly assigned to watch me made a strategic blunder; he turned his back for a moment. Doped up, I raced around the hospital in my wheelchair until I came upon a first class hospital corpsman folding towels. I noticed, even with all my medication, that he was laughing at me. I couldn't have cared less. I sat there in my wheelchair, rocking back and forth, and told him all about my life and what I was going to do when I got out of the Navy. Ultimately I ran out of energy, slept for a very long time, and was returned to my ship after the pain medication wore off.

Photo 10-3

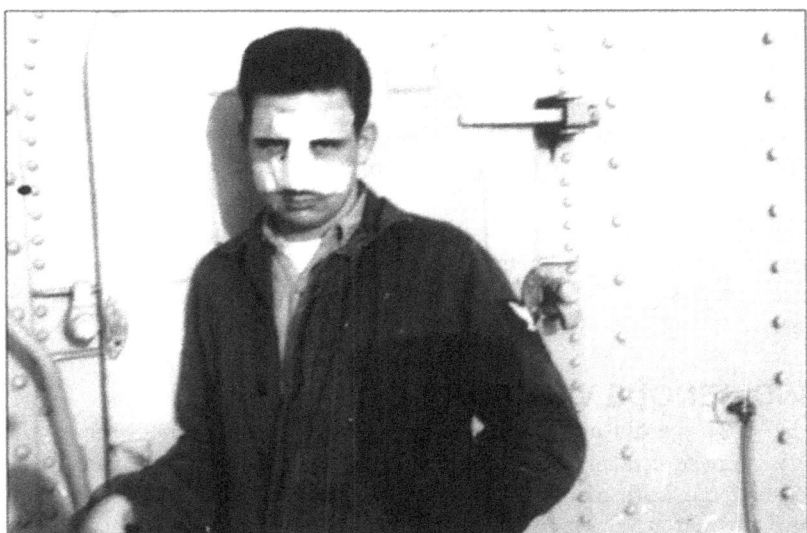

The author standing in front of a watertight door aboard ship; exhibiting facial damage from being on the losing side of an unprovoked assault in a bar.

My commanding officer wanted justice for the injuries to me, so the next day I was taken to *Lynde McCormick*. Her commanding officer had three crewmen assembled that he believed were the culprits who had beaten up Garvey and me. Asked if I could identify them, I looked at them and them at me. I really wanted someone to pay for rearranging my nose, but in good conscience I could not say for sure that they were the guys. So, I answered, "No sir." If they were the offenders, they were very lucky that day. Commanding officers tend to take care of their good sailors and look for ways to discharge from the Navy, or otherwise remove from their ship, habitual troublemakers.

Several days later, I went on a tour of Nagasaki with my face bandaged up: a molded piece of plastic and a metal brace over my nose with two wide bandages holding it in place. One bandage stretched across my forehead and the other down the bridge of my nose and over my cheeks. What a sight. As I passed Japanese children, they would look up at me, then their eyes would return straight ahead, as they continued walking without any change in expression. Once, when I turned to see if there had been any reaction, some were laughing. I was not offended, and happy that I could add to their day.

BACK ON THE GUNLINE

Shortly after, the *Hopewell* returned to the gunline, providing gunfire support for troops ashore. However, on one occasion, we had the opportunity to supply the Marines on the beach with steaks and freshly-baked bread from our galley ovens. They were so pleased they gave us several items captured from the Viet Cong: an old Springfield rifle; a gas mask made from Naugahyde vinyl fabric with a piece of cellophane used to cover the eyes, and gauze for a mouth piece; and a pair of sandals made from old jeep tires. Accompanying these gifts was a message, "Thank you for the steaks, God bless you for the bread!"

INDIGENOUS VIETNAMESE CRAFT

Although we did not work with them, we frequently encountered the Vietnamese coastal force, comprised of sailing and motor junks that carried out both inshore and offshore "junk patrols." These craft of the "brown water Navy" navigated scary, confined rivers and typically had eyes and shark's teeth painted on the bow of the boat for good luck—and to scare the Viet Cong. Occasionally junks would come alongside the *Hopewell* for fuel, water, or provisions. We would also give them boxes of candy bars from our ship's store, which was likely a welcome change from eating only fish and rice.

Photo 10-4

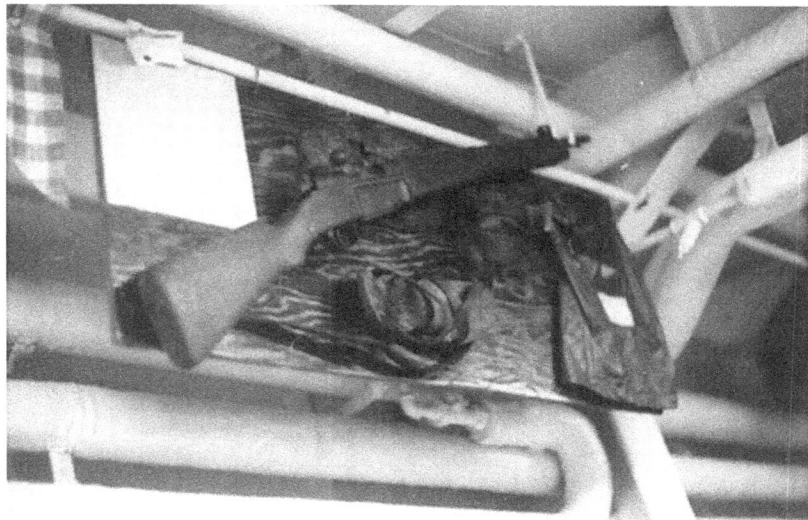

Captured Viet Cong items that Marines ashore gave to *Hopewell* in appreciation for the food we sent to them.
(Courtesy of former RD3 R. Mike Sohikian)

One of the junks had a washing machine strapped on its fantail to serve as the ship's laundry. I wondered how they kept it operating in the humid, rainy conditions in Vietnam that quickly rusted anything made of metal. I thought to myself that it would be one hell of an expense to have it repaired via a service call from Halldorson Appliance in Chico, California!

On one occasion, we were operating in the South China Sea with the attack carrier *Constellation* (CVA-64) and other units of her battle group, doing plane guarding, screen duty and other tasks. One night, either our conning officer gave an incorrect rudder order, or the helmsman put the wheel over to starboard instead of to port. In any case, *Hopewell* cut directly in front of the "Connie," which could have resulted in our being cut in half like the destroyer *Frank E. Evans* (DD-754) by the Australian carrier HMAS *Melbourne* on 3 June 1969. The captain was in his sea cabin, but was immediately aware of our situation. A rudder angle indicator and compass, mounted within the cabin, allowed him to monitor ship control course while off the bridge. As soon as he reached the bridge, he yelled "all ahead flank." By this action, *Hopewell*'s speed increased to maximum and we were able to get clear of the carrier's path. The North Vietnamese couldn't get us, but we could have been done-in by one of our own ships, a very big one sailors termed a "bird farm," aircraft being called birds.

Photo 10-5

Vietnamese Coastal Force craft flying Republic of Vietnam flag while underway. (Courtesy of former PN3 Johnny Junior Sharp)

LIVING CONDITIONS ABOARD SHIP

Prior to this deployment, our war time complement (the ship's crew size) had been increased. To accommodate this change, extra berthing was added throughout the ship including three "racks" (beds) in ship's office. The racks, stacked three high, were slung from one another with chains. The bottom one covered the top of our desk, mandating that it be "triced up" (hauled up and fastened) when not occupied to allow us to work at the desk. We had no rack lockers; we stored our personal clothing in standing lockers in a corner of the ship's office.

We worked and slept in the same space, "working at home" before it became fashionable. I did occasionally work in my skivvies after work hours but only behind a closed door.

A personnel man, Pete Petersen, who also worked in the ship's office, had a state-of-the-art reel-to-reel tape recorder. He received music from back home that included songs such as "Monday, Monday" by The Mamas and The Papas. The song was great. We listened to it over and over and over again. Another great song was "Cotton Candy" by Al Hirt, and of course anything performed by the Beatles. The Beatles album *Rubber Soul* had many notable tunes including "Norwegian Wood," "Nowhere Man," "Michelle" and "I'm looking Through You."

When the *Hopewell* was at sea, crew members that were not on watch at night were able to enjoy watching a movie on the messdecks. The movies made their rounds among Seventh Fleet ships via manila "high lines" (ropes) used to transfer material and/or personnel between two ships. Unfortunately, the messdecks, located two decks down in the innards of the ship, had only a few vents providing fresh air. Once a movie started, nearly everyone lit up cigarettes and the smoke was so thick you could hardly see the screen, your eyes burned, and coughing by many non-smokers, including me, ensued. While some might contend that the entertainment provided was worth this discomfort, I was not one of them. A word to the executive officer resulted in the "smoking lamp" being put out during movies.

TYPHOONS AND DIRTY SEAS

The dangers of the sea should always take precedence over the violence of the enemy.

—Rear-Admiral Ben Bryant CB, DSO and two bars, DSC

During the latter part of the deployment, the *Hopewell* encountered the tail end of a typhoon, which tossed us about like a cork in a washing machine. Everyone, from the captain on down, was seasick. The force of the waves ripped loose brackets which secured tanks of oxygen and acetylene to the bulkheads on the main deck; the bottles rolled around the deck, creating both missile and explosive hazards. One of the gun tubs on the 01 level was dented by the force of a particularly high wave.

The *Hopewell* had clinometers mounted on the bridge and in the combat information center to visually display how much she heeled over on each roll. We were in several large storms while I was aboard. The standard practice on such occasions was to turn the ship "bow into the seas." This was designed to prevent a vessel from broaching, and being trapped "beam to the seas" unable to maneuver to save herself. In such a case, each successive wave might roll the ship even farther over, until, unable to right herself, she would "turn turtle" and sink.

Sleeping during rough seas was a challenging proposition. I learned, like everyone else, to sleep with my hands clenched around the metal framework of my rack to avoid being thrown out of it. Eventually, you subconsciously only clenched when the ship took a rather large roll. One guy was transferred off the ship because he became so dehydrated from being seasick that he could keep neither food nor liquids down. One day at sea, a mess cook unwisely decided to dump garbage over the side of the ship in heavy seas. After arriving topside, a wave came up over the side and slammed him against a bulkhead knocking him out. His body then slid beneath the lifelines over the deck edge; the upper half was precariously perched on deck and his legs and a portion of his hips were hanging over the side. Someone on the 01 level above looked down and saw him, rushed down a ladder and pulled him to safety. The next roll of the ship could have launched him into the sea, where he would have died if no one would have sighted him, as he was unable to call out for help.

TAMER LIBERTY IN SUBIC

During one of our visits to Subic Bay, we had the opportunity to go swimming at Grande Island, known for its recreational opportunities. After going ashore from the *Hopewell*, we were transported by the "Cattle Car," a big gray trailer with lots of air holes like the type used for livestock. The air holes made the hot stuffy air and closely packed bodies almost bearable. Arriving at a beach near the island, we boarded a "Mike boat" (LCM amphibious landing craft) for ferrying to our destination. Once there, Johnny Sharp and I stripped down to our Navy-issue, dark blue bathing suits, which looked like baggy speedos. As we were preparing to go into the water, we noticed a guy with a spear gun, and struck up a conversation with him. We could tell he was a sailor because he had the same classy swimming suit as Johnny and I.

He was bragging about shooting at ten-foot sharks with his spear gun. Since his spear had frog gigs on the tips (pointy little tridents that would just annoy a shark), we did not believe him. And, if he was engaged in such activity, he was lucky to not be dinner for a mad target.

There was not much to see, the water being only about three or four feet deep; seaweed gently undulated about a foot below us as we swam. As I was swimming away from the shore, I turned around for some inexplicable reason, and spotted a shark maybe three feet long right behind me. At the time, it looked awfully big. Johnny was swimming ahead of me, so I stood up to await his head coming out of the water, then shouted, "SHARK!!!" We swam furiously—probably not a good idea, because our commotion in the water could possibly draw larger sharks—and were happy to gain the sanctuary offered by terra firma.

Later, I sighted an old high school classmate, whom I had not known was in the Navy. There he was smiling at me, good ol' Mike Behr. We both agreed that it was a small world and reminisced about the old times and our more recent experiences.

Although Johnny and I sought pristine waters in which to do our swimming, others were not so choosy. One moonlit night, as I was walking up the brow to return to *Hopewell* after liberty in Olongapo, I heard a commotion. Looking aft, I saw one of the guys from my division standing on the starboard screw guard. (This type device is designed to prevent a ship's propeller(s) from becoming damaged.) The officer of the deck (OOD), who would normally be positioned on the quarterdeck, was aft trying to coax the sailor back aboard the ship. The guy was crying and greatly agitated, and the OOD was having no luck in handling the situation.

After a short time, the captain and the executive officer returned aboard ship. They noticed what was going on and the commanding officer went to the fantail and talked at length with the unhappy junior enlisted, who was assigned collateral duties as a compartment cleaner and did not want to do that menial task anymore. The captain told him that he would look into the matter the next day and see what he could do for him if he would just come back aboard. The captain was successful, but as he was making his way up the starboard side of the ship to his cabin, over the ship's announcement system came, "Man overboard!"

The seaman had apparently changed his mind and had jumped into the putrid, filthy, oily, and feces-contaminated water of what we affectionately called "Pubic Bay." As he swam away from the ship, the *Hopewell*'s search light was directed on him. The captain ran to the port side rail and called to him on a bull horn, upon which the miscreant shifted to a back stroke and held up the widely recognized single digit sign of ill will. A motor whaleboat was dispatched to collect him and return him to the ship. Once pulled aboard the boat, he made it very clear by fighting with his rescuers that returning to the ship was the last thing he wanted to do. For this, he was hit in the mouth. While a ship's commanding officer assigns punishment during formal procedures, after an event has been thoroughly investigated, sailors sometimes engage in "direct feedback" to inform one of their own that their actions are not acceptable.

PINUP GIRLS

Many young men, including sailors, have aspirations of being a hero. I wanted to be an international playboy. Of course, I did not realize that great looks, sophistication, and a large bank account, were some of the prerequisites one would likely need, none of which I had. What I and other young sailors did have were photos of "pinups," usually safeguarded inside a locker, where they could be admired as one accessed one's few possessions aboard ship. The servicemen of World War II had Betty Grable and Rita Hayworth to remind them of home; those of the Korean War, Marilyn Monroe and Jane Russell. We that served in Vietnam had Joey Heatherton, Ann-Margaret, and eventually, Carol Doda. Six of these ladies were extremely attractive actresses, who were also able to sing and dance. The non-actress, Carol Doda, was admired by men for other different, but obvious attributes.

As I was reading the Armed Forces newspaper *Stars & Stripes* one day at sea, I noticed an article advertising that any serviceman who mailed Carol Doda his name and address would receive an 8 x 10 glossy photograph of her. I had never heard of this dancer but became intrigued, as you can imagine. Further investigation revealed that Miss Doda was a topless dancer at the Condor Club in San Francisco, famous for her double "Ds" or triple "Ds" dubbed "twin 44s." Some also referred to her attributes as "San Francisco's Twin Peaks." I, being a healthy red-blooded American male, sent her my name and the address of my duty station, and then waited and waited, and waited some more, for a manila envelope with a glorious depiction of womanhood at its best. As mail calls aboard ship came and went the excitement and anticipation grew. But alas, the envelope never came.

Many years later, I was driving home from San Francisco and listening to the radio, I heard Carol Doda being interviewed by a talk show host. They invited callers with questions and I could not resist, so I got on my cell phone and called. (This was before it was illegal to use your cell phone while driving.) She apologized for my not getting the picture and said that if I sent her my address that I would get a copy of that long-awaited photograph. I did not pursue this option. By then too many years had passed for the same magic to be evoked in me by her likeness.

Playboy magazine fueled many dreams that never came to fruition. There was one depiction of a debonair guy in a maroon smoking jacket driving an MGTD (a sports car) painted British racing green. I believed I too could be a dapper gentleman of the world and get a plethora of rich and beautiful women. Never mind that I did not have thick, black, wavy hair, or blue eyes (for that matter), was not ruggedly handsome, and had no money. The stimulus for a more realistic fantasy came from standing next to the ship's forward stack at night enjoying the warm air being vented from the ship's engine room, as the destroyer cut through Pacific waters. I would stand there—with my hands behind me on cold nights to warm them, and look out over the ocean as waves slapped against the ship's hull and moonlight bounced off undulating waves—enjoying the intoxicating experience.

So, what did I think about while standing there enjoying the solitude? Suspecting that I would never be an international playboy, all I could think about was, when I got home to Chico, getting a six-pack of beer and heading for the Bear Hole or Salmon Hole. These were the names of pristine swimming spots on a creek that ran through Bidwell Park. Bidwell Park, one of the largest municipal parks in the United States, encompasses miles of land stretching from the foothills into the heart of Chico. These spots along the creek were framed by volcanic lava rocks that channeled the crystal clear water rushing down from the Cascade mountain range. I would dream of sunning myself on one of the big rocks with a trusty six-pack of Olympia, getting drunk and acting silly. If luck were with me, there would be unencumbered chicks present, and one would go home with me. Such are the dreams of a homesick sailor.

102 Chapter 10

Photo 10-6

The "Bear Hole" in upper Bidwell Park, Chico, California.

LIBERTY IN HONG KONG

Hong Kong had been called the most beautiful city in the world. For sailors, liberty activities spanned the spectrum from swilling alcohol in a "watering hole" to visiting exotic locations in and around the city, which were themselves intoxicating, as were the women. It was possible to go from incredible wealth and prosperity to abject poverty in just a few blocks. One of Hong Kong's attractions was the Tiger Balm Gardens. Accenting the gardens was a large pagoda and statues painted with garish colors. One, the Mahatma Gandhi, was surrounded by beautiful fresh flowers. The tranquil experience the gardens offered brought me back many times.

To get around the city, one could engage the services of a rickshaw, like hiring a cab in other cities. One of the pastimes enjoyed by sailors visiting Hong Kong was rickshaw races. For a few Hong Kong dollars (a Hong Kong dollar was then worth about eighteen cents) you could race your buddy. "May the best man win," meant the Chinese guy pulling the rickshaw, not the "anchor clankers" sucking on a beer. The rickshaw drivers were in great shape, not an ounce of fat on them. Fred McClure and I had a race one night and I think his guy beat my guy or, maybe I won, who knows and who cares. My memory is a little dim, but we probably engaged in these antics, after tipping a few, and were accommodated by either very patient or bored rickshaw drivers.

While in Hong Kong, I took a tour to Kowloon and the New Territories, which went through the Forbidden City and other hamlets that dotted the countryside to the frontier of Red China. During a stop, we were told, while gazing out over a vast valley, that we were looking at Communist China. Prior to boarding the bus, we encountered an old Chinese man who had the stereotypical Chinese beard and Fu Manchu mustache. On his head was what we called a "ping pong" hat, one that is round and comes to a point in the middle. I indicated that I wanted to take a photograph of him and he agreed. As soon as I took the picture, he held out his hand and demanded a dollar. I hadn't anticipated that requirement, but I didn't mind.

One might think that with all these type of opportunities, sailors would not find ways to get into serious trouble, except, perhaps, that involving women. One would be wrong. I was shocked to learn that one of our firemen, a ship's engineer, was arrested by Hong Kong police for getting really drunk and stealing a British lorry (truck). Stealing it was bad, but not as bad as ramming nine other lorries with it. Unlike large Navy ships, such as aircraft carriers, destroyers do not have a brig to house particularly grievous offenders awaiting "Captain's Mast." I am not sure what steps our officers took to get him back from the civil authorities, or what transpired during such procedures, but once aboard he was locked up in a steel mesh cage, in the after engine room. This cage was used to store tools, preventing theft and thus ensuring they were available when needed, but offered sufficient additional space to temporarily house a miscreant.

He was court-martialed, charged with destruction of property and bringing disgrace to the Naval Service. I'm not sure which jurisdiction took precedence, the Hong Kong authorities or the Navy, but I think that both had their way with him.

SHUTTERBUG

Photo 10-7

F-4 Phantom catapulted from the USS *Constellation* (CVA-64) in January 1969 for a strike mission over Vietnam.
Naval History & Heritage Command Photograph #L01-20.08.01

One day the *Hopewell* was engaged in plane guard duty, which involved maintaining position behind a carrier in case a need arose for pilot rescue. Landing an aircraft on a carrier, in comparison to taxiing to a stop at the end of a lengthy airstrip ashore, was dangerous, particularly at night, and things did not always go right. Navy aircraft continuously flew right over us on approach to her flight deck. This gave me the opportunity to take incredible pictures of the bellies of F-4 Phantoms and A-6 Intruders. I saw myself as a modern-day Ansell Adams creating more dramatic photos. While thus engaged, many of my shipmates snickered as they strolled by. Eventually a guy who stopped to watch me take three more pictures took pity on me and said, "Your lens cover is on."

Photo 10-8

A-6A Intruder aboard the carrier USS *Forrestal* (CVA-59) on 25 August 1963.
U.S. National Archives Photograph #USN 1080873

I never did get the photographs I sought. I was out of film. I eventually purchased a single-lens reflex camera which, unlike the double-lens reflex camera I had been using, allowed me to view what the camera was seeing. Thus, I would have known I had not removed the lens cap. Never again was I in a position to film Navy jet fighters passing so closely above me. The planes pictured above are similar to those planes I was attempting to photograph as they flew over us while being launched or recovered aboard the carrier.

However, all was not lost. I had also bought an 8mm movie camera and I took hours of film of the sea and Philippine Islands—that you could barely see in the distance—as *Hopewell* transited the San Bernardino Straits. Did I devote any film to the guys on my ship? No, because I then believed that pictures should be taken of things and places and that people only cluttered them. I was so wrong, and would treasure more pictorial memories of my shipmates, today.

A HERO REPORTS ABOARD

During the deployment, a new Hospital Corpsman reported aboard, and I went to sick bay to meet him and welcome him to the *Hopewell*. He was a big guy about six-two with blond hair and blue eyes, who looked like he just came from an Iowa cornfield. He was pleasant and mild-mannered and we became friends. After hearing that he had been "in country" in Vietnam, I asked him about it, but he had little to say. Being curious, and a yeoman having access to his service jacket, I reviewed his service. To my pleasant surprise, I found out that he had been awarded the Bronze Star Medal for performing a battlefield tracheotomy on a wounded Marine as bullets flew all around him. That Marine was alive because of his valor. I later asked him how he could perform this extraordinary task while being exposed to enemy fire, to which he replied, "It was my job."

I reflected back on the time I had picked up the 3-inch projectile dropped at the base of Mount 32. It had slipped out of the hands of the first loader, hit the gun mount deck, and then rolled and fell down to our deck. Though terrified, I picked up the gun round, ran to the side of the ship and threw it into the sea. The point-detonation round might have exploded, damaging the mount and killing the mount crew. Did I consider this action part of my job? Yes, I did. So, I suppose I also had followed the culture of the Navy. The enlisted boat captain of a 57-foot wooden minesweeping boat charged with keeping the Long Tau River in South Vietnam open to Allied shipping epitomized this philosophy. Interviewed, following the loss of boats and men on the river, he stated, "We know our job, and we will do it."

We also, like probably every other Navy ship, had one or more nefarious characters aboard, some much easier to identify than others. The *Hopewell*'s Postal Clerk was a great guy to go on liberty with. He was very personable and because he would always treat his shipmates to drinks and food, we naturally thought that he was from a wealthy family. It turns out that the "mailman" had been issuing U.S. Postal money orders to himself and then cashing them. Before we knew any of this, he went A.W.O.L. (absent without leave), apparently believing the authorities were closing in on him. Following his disappearance, I was assigned to type up a DD553, Department of Defense form. This document was sent to everybody who might have known him. It included his name, civilian address, description (height, weight, etc.), with a request to apprehend and contact authorities if his whereabouts were known.

The DD553 must have worked because our missing shipmate was caught, arrested and held in an Air Force stockade in Washington

State. A few months later, we were informed that he had escaped, and were instructed to put out DD553s again. This time it didn't work. They could not locate him. His body was found the following spring, only one hundred yards from the stockade. He had died of exposure. For every bad apple that finds his way into the Navy, there are hundreds of good guys, and a few real heroes.

Another hero soon reported aboard, Lt. Comdr. James Burpo, from "Swift Boat" duty with the Riverine Force. Burpo was my third executive officer, and also a recipient of the Bronze Star Medal.

VISIT TO MANILA

During one of *Hopewell*'s many visits to Subic Bay, I visited Manila, capital city of the Philippines. On the way to Manila, the bus tour I was a part of stopped at the resort town of Baguio, where the Japanese Army had sent soldiers for rest and relaxation during World War II. After arriving at Baguio, we were guided in canoes upstream to view the picturesque Pagsanjan Falls. There were water buffalo amidst the palm trees and lush green foliage that lined the banks. At the places where the river slowed to a trickle in shallows, we would disembark and our guides would carry the boats to navigable water.

While shopping in Baguio following the river excursion, I received three offers of marriage. Boy, was I a hot commodity. Some of the guys warned me that the girls just wanted to get to the States, but I knew better; they had fallen victim to the "Halldorson Charm!" Such occurrences were commonplace, and reflected the desire/desperation of young women to escape the poverty and limited opportunities that existed in the Philippines and to make a better life for themselves.

One example was the bar girls in Olongapo, who participated in different gimmicks to attract sailors and keep them coming back. One establishment boasted a version of musical chairs and gave a prize to the winner. Many, if not most, of these girls had no choice regarding such employment. The Philippines was a poor country and commonly when word reached outlying areas that Navy or civilian ships from the United States or other countries were expected, attractive girls from rural areas were sent to the cities to be hookers. As soon as possible, I removed myself from this situation.

As we traveled by bus across rural Luzon (the largest island in the Philippines), on Saint John the Baptist Day, 24 June, we were greeted by little Filipino kids adhering to an age-old tradition which was part of their religion, baptizing buses and cars by throwing water on them. Unfortunately, the windows of our bus were open due to the heat and humidity. We were wearing white uniforms, and the little urchins were obtaining their water from mud puddles on the side of the road. Eventually the bus came upon a roadside make-shift shack that sold items to eat and drink, and stopped. I bought a Baby Ruth candy bar and was happy with my purchase until, while talking with my seatmate, I happened to glance down at the uneaten portion in my hand. Wiggling from what turned out to be a "protein bar" were little white worms, yuck!

After hours and hours of bus travel, we finally arrived in Manila and drove past the Presidential Palace, the Philippine equivalent of the White House, where Ferdinand and Imelda Marcos lived and ruled. The streets of the city were crammed with jitneys (World War II era jeeps, painted in garish colors, carrying passengers for a low fare). Blaring from radios everywhere were the sounds of, the Beatles and other British Invasion bands. We admired old churches and cathedrals that were very ornate and beautiful, and fortifications dating back to the Spanish Colonial era, which ended with the outbreak of the Spanish–American War. However, the most compelling sight—and one always on a sailor's mind—were beautiful girls.

THE "ARNHEITER INCIDENT"

Much of this narrative about my second deployment has focused on time spent "on the beach" on liberty, partially because such memories come readily to mind, and because I discussed combat operations in depth in the chapter associated with the first deployment. However, we did, during this tour of combat off South Vietnam, operate at the same time as the destroyer escort USS *Vance* (DER-387). One of the most infamous events of the Vietnam War was the "Arnheiter Incident" involving a series of events and atmosphere aboard the *Vance* that led to the "relief for cause" of the commanding officer. This affair is aptly chronicled in the book *The Arnheiter Affair* by Neil Sheehan. Those not familiar with this embarrassing episode in U.S. Naval History should read about it. An officer quoted in a *Time* magazine article stated, "We all have a little of Captain Queeg in us, but Arnheiter had more than his share." (Captain Queeg is the central figure in Herman Wouk's best-selling 1951 novel *The Caine Mutiny*, which won the Pulitzer Prize for Fiction.)

Capt. Donald F. Milligan, commander Escort Squadron 7, relieved Lt. Comdr. Marcus A. Arnheiter of his command on 31 March 1966, while the *Vance* was refitting at Manila. Arnheiter had been aboard her for a mere ninety-nine days. Milligan had been commanding officer of the *Hopewell* during my first cruise, and had left her for new duty as a squadron commander. Milligan himself took command of the *Vance* until a suitable relief could be identified and report aboard.

Photo 10-9

USS *Vance* (DER-387) under way off Oahu, Hawaii, on 18 January 1968. Official U.S. Navy Photograph #NH 107602, from the collections of the Naval History and Heritage Command

In July 1966, following months of duty in the Far East and off Vietnam, the *Hopewell* was relieved of her duties and began the long Pacific crossing to return to San Diego. During the deployment, we had visited Yokosuka and Sasebo, Japan; Chimu Wan, Okinawa; Hong Kong; and Subic Bay, Republic of the Philippines. (Chimu Wan was known to servicemen as Buckner Bay in honor of Lt. Gen. Simon B. Buckner Jr., USA, who had been killed in action on Okinawa on 18 June 1945, during World War II.) While at Subic Bay, *Hopewell* sailors had donated their time and talents to aid a grammar school in Barrio Culis, Bataan, by painting the three buildings of the school and distributing books and clothes to the townspeople.

11

Ship's Office/Captain's Mast

Choose a job you love, and you will never have to work a day in your life.

—Confucius

Don't be afraid of the storms; be afraid of the ship and the captain! Forget about the outside factors, what matters is the internal power!

—Mehmet Murat Ildan

A majority of this book has been devoted to ship operations, and adventures of a young sailor ashore. In truth, practically all of my time was spent in the ship's office carrying out my duties. Like other areas of the ship, we had a combination of very capable, mostly adequate, and a few less than desirable individuals. The Navy did not care for the latter type of sailors. For almost all my time aboard *Hopewell*, Senior Chief Taitano ran the ship's office. He was a very, very good yeoman and a 4.0 sailor, one who exemplified the finest the Navy has to offer. He was Guamanian and when the ship once visited Agana, Guam, his entire family came aboard for a visit. He was superbly efficient at his job and unwaveringly professional. He prided himself on knowing where every piece of correspondence, directive or mail was aboard ship at any given time. He accomplished this via transmittal forms which he attached to every piece of paper in his clerical kingdom.

The transmittal form had places where the officers, department heads and also the captain and the executive officer—each person who received a piece of mail or official correspondence—would initial to acknowledge receipt. The chief kept copies of transmittal forms and thus knew at any given time who had read a piece of mail and who had returned it to ship's office. If not returned, he knew who had it. The system worked flawlessly.

As previously noted, during one visit to the Philippines, I had the opportunity to go on a field trip to Manila. A prerequisite to this opportunity was my completing all yeoman work in my "in basket." I filed papers until an hour past midnight when I determined that I could not complete my tasking, so I did the unthinkable. I stuffed the remaining burden residing in my "in basket" somewhere else and later that morning I went to Manila. I was not smart enough to realize, or did not care, that this action would not go unnoticed by the chief.

I was on cloud nine when we returned late that night to the *Hopewell* at Subic Bay. Upon stopping by the ship's office I was told by the duty yeoman that I was "on report" because the chief was very mad at me. More accurately, he was royally pissed. It seems that during my absence he needed one of the letters that I had not filed, and failed to locate it where it should have been. He finally found it in my "in basket," the contents of which were supposed to be completely filed. Chief Taitano threatened me for about a month with the "report chit" he had filled out detailing my offense, but did not submit it to the command. He would periodically push that piece of paper into my face and say, "Halldorson, you F***k Up!" His face would get twisted and contorted, and beet red. With his eyes bugged out, he looked like a stroke waiting to happen. He eventually tore up the chit, which was then unnecessary, as I had learned my lesson.

Much later, during my second cruise aboard the *Hopewell*, the senior chief transferred off the ship to a new command leaving, by default, Yeoman Third Sharkey and myself in charge. We promptly decided that we did not need to use the transmittal forms. To us filling out and maintaining them was a waste of our precious time, which could be better used drinking coffee or yakking with each other instead of using the old—archaic—methodology. We were of the modern Navy. The first time the captain asked for a particular piece of correspondence and we could not produce it, we recognized that our former boss had been a "paper god." Like parents, he had gotten a lot smarter with the passage of time.

For a relatively short period, and for good reason, a Yeoman First was in charge of the *Hopewell*'s ship's office. He had been stationed in country in Vietnam, and reported aboard during our second deployment. When the ship was in port, we would come into ship's office most mornings and find him slumped over his typewriter with his head on his arms, reeking of alcohol. He had been a Chief Yeoman in Vietnam, been busted to First Class, and was then dumped on the *Hopewell*. Apparently his demotion was for losing a jeep and a .45 caliber pistol issued to him. We suspected he had lost them while drunk. In any case he was a broken man who had lost something more important—his pride, as well as the respect of his subordinates and shipmates.

Photo 11-1

Yeoman Third Halldorson at work in *Hopewell*'s ship's office.

SHIP'S NEWSPAPER

Aboard ship, the locally produced equivalent of a newspaper is the POD (Plan of the Day), which disseminated information to the crew. Like any newspaper this publication required the efforts of an editor and a copywriter, etc. The XO (executive officer) would hand write notes covering such things as the uniform of the day (for working and meals), ship's schedule, time of sunrise and sunset, and everything in between. When the *Hopewell* was overseas, he might include a warning to the crew about ladies of the evening and associated diseases, local customs, and general warnings about expected proper behavior. The

XO would call the yeoman to his stateroom to get the POD notes, then the yeoman would take the notes to ship's office and type them up on a blue waxed paper form for the mimeograph machine. The typed POD would go back to the XO for proofreading. Since the POD was NEVER correct with the first typing, the XO would make his changes and the yeoman would reflect them in a corrected copy. After finally satisfying the XO and obtaining his signature, the yeoman would run off copies on the mimeograph machine. When they were dry, he collated and stapled them, and distributed them throughout the ship.

On newer destroyers you could walk the entire length of the ship via interior passageways. Not aboard our beloved *Hopewell*. As you proceeded fore and aft, you had to, at times, venture outside the "skin of the ship" onto the weather decks. All those times to the XO's office, and the distribution of the POD, meant many trips up to the 01-level. Often, this was an opportunity to enjoy blue skies, but in heavy weather, wind and rain whipped into you, and you had to pull yourself along the lifelines hand-over-hand. Neither snow nor rain nor heat nor gloom of night stayed the duty yeoman from the swift completion of his appointed rounds.

GOOD ORDER AND DISCIPLINE

Being a yeoman, I was responsible for recording the results of non-judicial punishment, awarded by the commanding officer at "Captain's Mast," to violators of rules and regulations. Among the most common offenses were such things as being late to morning muster or disrespect to a senior petty officer. I attended many of these solemn affairs for various violations of the UCMJ (Uniform Code of Military Justice). It was my duty to type up the charges against the sailor in question, and then to give them to the captain. During "mast procedures," the accused would stand at attention in front of the captain; I would stand beside the captain and record the results.

Some of my shipmates accumulated quite a "rap sheet" of infractions during their time aboard ship. The captain would view all the priors for that particular individual, ask his chief petty officer and division officer to comment on the quality of that person's work, then dispense justice. There was normally some leniency for minor offenses, if the sailor had not been to mast before, and consideration would be given if he was a good performer. Repeat offenders and/or slackers could expect no sympathy.

I was also involved with administrative matters pertaining to court-martials. One involved a seaman accused of stealing money left on a bunk in the Operations/Communications compartment. Someone saw him take the money and reported it. He claimed that he found the money on the bunk and was going to turn it into ship's office so that it would get back to its rightful owner. Thieves are particularly detested aboard ship, and the captain ordered that a Special Court-martial be convened. The court-marital was conducted while the ship was in port at San Diego. Per naval tradition, a table in the officer's wardroom was covered with green cloth. Sitting square in the middle of the table was a brown reel-to-reel tape recorder, in an alligator-hide covered box with one little microphone to record the proceedings.

Normally in such cases, the accused is represented by a Navy-provided Judge Advocate General lawyer. In this case, the seaman's parents obtained the services of a civilian attorney instead of accepting those of a Navy JAG. The civilian lawyer brought in letters from the accused's pastor, his scout master, and other prominent people in his life who could vouch for his character. It turned out that he was an Eagle Scout. He was acquitted of the charges and shortly after, we put to sea.

I typed up the transcript of the court-martial while sitting cross-legged on the deck intently listening to that World-War II vintage tape recorder. The ship was rolling in heavy seas, and I could not otherwise maintain my balance. In transcribing a court-martial, you must type every "ahhh" and every "ahem." You cannot leave anything out. I had to rewind the reel to start over again in order to hear every faint nuance. My other materials were an old Olivetti typewriter, a stack of legal-sized paper and carbon paper, and a typewriter eraser. Each time I typed a wrong letter, I would have to stop, twist the roller up to the appropriate spot, erase the bad letter five times (one on the original document and four on carbon copies) and re-type it. It took me many hours to produce the final document—over sixty legal sized pages.

12

Return to San Diego/Overhaul

There's nothing more beautiful than the way the ocean refuses to stop kissing the shoreline, no matter how many times it's sent away.
—Sarah Kay

Our memories of the ocean will linger on, long after our footprints in the sand are gone.
—Unknown Author

After arriving in San Diego on 1 August, the *Hopewell* had a busy schedule ahead of her. Activities included serving as a school ship, first, for attendees of Navy gunnery and then, anti-submarine warfare schools. In late September she served as host ship for thirty Naval Reservists during weekend duty.

However, local operations left time for the ship's crew to enjoy themselves ashore. One day in August, a bunch of us decided to go to Mission Beach in San Diego, drink beer, and try to meet girls. We got the beer and stretched out on the sand to await any possibilities that might come our way. We knew that we could not get sunburned because it was overcast—no sun, no sunburn, right? One of the popular songs then being played on the radio was "Sunshine Superman" by Donovan. Boy, did I love that song; we were sunshine supermen on a beautiful beach, away from the ship. We drank, listened to music and watched girls.

Late in the day, as it turned a little cool, one of the guys found some cardboard boxes behind a nearby supermarket to start a driftwood bonfire. Later, as the fire burned down, one of us threw a brown paper bag on it to keep it going. As it burned away, we saw the pants and belt of one of our group members inside the bag. But alas, it was too late to save it. Wanting to extend our day of leisure, we went to a movie that night. The weather was starting to be unseasonably cool, the theater was air-conditioned, and I was cold.

The next morning we shifted from wearing dungarees to undress blues aboard ship. The latter uniform, being of wool material, was warmer than dungarees, but also less comfortable and scratchy. Thus attired, I moved very slowly, being so severely sunburned that skin was peeling off my back and legs. This discomfort gave proof to the old adage "moderation is the key," and taught me that the sun's rays pierce clouds even when the sky is almost completely overcast.

SHORT-TIMER

After a few months in San Diego, the *Hopewell* was sent to the Hunters Point Shipyard in South San Francisco, for an overhaul period. During this time, I became a "short-timer," meaning that I had less than thirty days remaining of my enlistment. The ship's "retention officer," Electrician's Mate Chief Schneekloth, approached me with a proposal to "ship over," to re-enlist and stay in the Navy. After hearing about the benefits of "shipping over," I must admit that I was attracted by the $5,000.00 bonus.

For a minute or so I thought about the '55 Chevy that I had seen on a car lot in Compton, California, and would have loved to purchase. However, I had aspirations to experience life outside the Navy. I declined to remain a sailor, and asked him not to be offended. This proved to be a good decision.

OVERHAUL IN DRYDOCK

A preparatory step to entering Hunters Point Naval Shipyard was to offload all our ammunition at the Concord Naval Weapons Station, an annex to the Mare Island Naval Shipyard, at Vallejo. Handling ammo by hand was both a strenuous and dangerous activity. To ensure the crew truly understood the risks associated with mishandling ammunition, the Navy briefed us on the procedure and showed us a photograph. A picture *is* worth a thousand words. This one showed a destroyer aboard which a crewmember had dropped a hedgehog, an explosive anti-submarine weapon. The resultant detonation had blasted the ship's bridge back onto the forward stack and tragically, scores of men lost their lives. We were very careful off-loading our ammunition.

Seeing the *Hopewell* out of water and up on cement keel blocks after docking was quite a sight. The first night aboard our drydocked ship was miserable. The *Hopewell* was "cold iron," meaning the engineering plant was secured. Without fire in one or more of her boilers, she could not make steam—the source of heat to the ship—and there was no "shore steam" available from the yard. These conditions were made worse by winds whipping through the dock on a particularly cold Bay Area night. It was definitely a "Three Dog Night," but no dogs were available to huddle with. The next night we were relocated to temporary barracks ashore with heavenly heat and lengthy hot showers. The conditions aboard ship were, however, not our greatest challenge.

Hunters Point was infamous for being a particularly dangerous neighborhood. While the *Hopewell* was in the yard in the fall of 1966, a sniper was on the loose, and people in the area were frightened of being his next victim. Standing at a bus stop with some other guys, the conversation turned to us all feeling that we had cross-hairs on us. This may sound crazy, but the feeling was real. Wariness permeated the area, as it did years later, when the Zodiac Killer was on the prowl. If the killer was ever caught, it occurred after we had left the yard.

Of course, most sailors can make the best of any situation. I was invited to a party being hosted by the ship's company of another ship in the yard. Considering it "bad form" to turn down a free drink, I strolled down the pier that night to the building where the party was being held. I was pleasantly surprised by the amount of food and drink—beer, wine, and harder stuff. I ate too little and drank way too much before making my way back to our barracks and hitting my rack.

The next morning I had the mother of all headaches and my stomach hurt, I could not get up and was really hung over. Sighting two of my shipmates across the barracks, I asked them what happened the night before. "You don't know?" they replied with astonishment. It seems that I had staggered into the barracks and sat down on someone else's rack. The occupant offered me a bag of peanuts likely believing I could use some food, which I promptly poured over his head. I then jumped up and ran across the room with my head down, and straight into my own locker. That explained why my head hurt so much, which I had attributed solely to too much drink. Three guys picked me up and put me in my rack. I swore I would never drink again, but this would prove to be a short-lived vow.

After a weekend liberty at home, I found myself at the Greyhound bus station, on Wall Street in downtown Chico, waiting for the bus to San Francisco. After boarding, I found a seat next to a rather engaging young lady named Nancy. She and I shared a very enjoyable ride to San Francisco where she was attending a business college. We talked and laughed together. I later called her several times. She was always very pleasant and it seemed we had a lot in common. She finally informed me that she had a boyfriend in the Air Force. Having no luck with girls on leave and liberty, I devoted the remaining time I had left in the Navy to my duties aboard *Hopewell*.

13

Return to College Life and Duty in the Naval Reserve

I wanna go to a party school! Yeah, Chico State!

—Originator of statement unknown; sentiment very common

I was granted an early-out, allowed to leave active duty a little earlier than my actual date of separation, to enable me to return to Chico State College for the Spring Semester 1967. Dr. J. Russell Morris, a professor of education, had written a letter to Chico State on my behalf and was able to get me reinstated. After returning to student life, I affiliated with the Naval Reserve, my total six-year commitment not yet having been fulfilled. In coming years I, like many other weekend warriors, would periodically find myself yearning to be aboard a ship at sea while forgetting the stark, spare life that accompanied such endeavors.

Sailors usually referred to their peers as "short-timers" if they had less than a year remaining on their enlistment. Once you were down to thirty days, you were really "short," and the remaining days might be scratched off a short-timer's calendar as they expired. On the *Hopewell*, this accounting was done on a sheet of paper with the outline of a naked girl divided into sections numbered from one to thirty. When the calendar was completely filled in, your enlistment was ostensibly up, unless you were extended for the "good of the Naval Service," which sometimes happened during war.

The *Hopewell* transferred me to "separations" at Treasure Island in San Francisco Bay to be released from active duty. The officer in charge of supervising discharges decided that I was to remain there an extra three weeks, typing discharge papers. His staff must have been short of clerical rates. I manned a trusty typewriter filling out discharge papers for the guys that left the ship with me and many others from other commands. I had some lighter moments, though. One sailor was a Native American with the last name of Screaming Eagle. However, the surname that took the prize was a Hospital Corpsman named Sick.

Eventually, my time came and I boarded a Greyhound bus in San Francisco for Chico. I sat in the rear of the bus next to a window so that as I watched the landscape go by, I could snooze, relish my freedom, and fantasize what college would now be like. It turned out I was on the "milk run." The bus stopped at every small burg from San Francisco northward. But that was okay, I was on my way home.

My reverie was soon interrupted when a very large red-headed woman, who was adorned with tattoos and had questionable hygiene, sat down next to me. She wanted to talk, so I sat there and I listened to her life story, which included her having been in prison. I wanted to be left alone and think about my impending return home after war duty, but did not tell her this. Before I got off in Chico, she told me her name and gave me her address; I did not pursue the invitation to stay in touch.

COLLEGE LIFE AT CHICO STATE

I must study politics and war that my sons may have liberty to study mathematics and philosophy. My sons ought to study mathematics and philosophy, geography, natural history, naval architecture, navigation, commerce, and agriculture, in order to give their children a right to study painting, poetry, music, architecture, statuary, tapestry, and porcelain.

—John Adams

After returning to Chico, I found that I had changed, and so had the town. The girl that I had left behind was gone and so was my car. I had grown-up and so had Chico. My beautiful brick high school was soon torn down following a major earthquake in Southern California due to fears that it might not survive a major tremor and was unsafe. Chico Senior High School was a gorgeous brick building with white marble gables and archways. Its large sweeping driveway led in from the Esplanade, the thoroughfare that ran in front of the school. In fact, the contractor hired to demolish it had a difficult time doing so with a wrecking ball. That proved the building had been the victim of an increasingly "safety conscious" society. Significantly, there are no major earthquake faults near Chico, and the school had stood proudly, unharmed by nature since 1920.

Two other major projects had occurred in my absence: the construction of the world's largest earth-filled dam across the Feather River running through the nearby city of Oroville, and a freeway over Bidwell Park. Bidwell Park was the backdrop for the 1938 motion picture *Robin Hood*, starring Errol Flynn and Olivia de Havilland, filmed in Chico. This beautiful park was then the largest municipal park in the United States. Today, Bidwell Park is still the largest in America for a community the size of Chico. My hometown's other crown jewels are Chico State University, the second oldest public state college in California, and Sierra Nevada Brewery, one of the premier craft breweries in the United States.

One of my first priorities upon returning to college was to learn how to do all the new dances with all the bitchin' moves that had come out while I was overseas. This would also be, I believed, a good way to get a girl. While registering for classes, I signed up for a physical education class that offered instruction on all the latest dances. I was feeling pretty good about this decision while walking back across campus. I was going to learn to do "The Swim," "The Monkey," and my all-time favorite "The Jerk."

Photo 13-1

I took a class at Chico State to learn to "groove to" rock & roll music at party scenes, such as this one.

It just so happened that I was walking behind two girls who could not contain their glee at the possibility of a single guy in their dance class; the one for which I just signed up. As they continued their discussion I learned they wanted to see this guy in his leotards. Yikes, I had signed up for "art" dancing, or modern dance. I never made it to that class. I don't recall whether I dropped it, or just took an incomplete or "F" grade. All I know is that I never danced in tights. I had better luck with my other classes, and eventually graduated in 1969 with a degree in Art. My love of this discipline led me to later pursue graduate studies in the same area and to become a professional artist.

Photo 13-2

Linocut print of *Hopewell* that I did at Chico State College in spring 1967. (In the artwork, "7th Fleet" and "VC" refer to the U.S. Seventh Fleet and the Viet Cong, respectively.)

CONCURRENT DUTY AS A NAVAL RESERVIST

While attending Chico State, I was also a member of the Naval Reserve, a "weekend warrior." One of my duties was being in charge of my unit's color guard. On Veterans Day in 1970, I marched the color guard to the free speech area of the campus. The girl I was dating, being liberal, did not like my being on campus in my dress blue uniform. I was likely a real embarrassment to her in front of her like-minded friends.

As we marched to the designated area, a World War I veteran and fellow member of the local VFW post, was in the audience observing the proceedings. Old Gus took exception to the way I was commanding the color guard and barked out an order. I guess I was not doing it the way the "dough boys" did it. In response to his command, two members of the guard broke right for an instant, and then rejoined the remainder of us who had continued straight ahead.

Afterward, I went with my girlfriend to the campus student union, still in my uniform and with a flush of crimson on my face. She took this opportunity to tell me, in front of others, that the United States "had it coming" at Pearl Harbor. We were not destined to be together after that.

Old Gus later sent me a letter of apology, expressing that he had only done what he deemed right at the time. His gesture was unnecessary; I was happy to have invoked the interest of a veteran of his war in the activities of our color guard on Veterans Day.

ANOTHER HALLDORSON "SHIPS OUT"

In October 1968, my brother Alan joined the Navy. After completing boot camp he reported to the tank landing ship *Caddo Parish* (LST-515), a unit of the Mobile Riverine Force in Vietnam. To get to his ship he was flown to the Tan Son Nhut airbase and from there made his way to his ship at Vung Tau. The *Caddo Parish*, an old World War II vintage ship, was employed as a floating base for "Sea Wolf" helicopters. She operated on various rivers, including the Mekong. During his service aboard her, the ship took Viet Cong machine gun, small arms, and RPG (rocket-propelled grenade) fire.

After the Navy put the *Caddo Parish* out of service, Alan was transferred to the *Iredell County* (LST-839), a newer amphibious ship assigned to Landing Ship Squadron Two. His GQ stations included being assigned to a damage-control party and as a .50-caliber machine gunner for a mount in an exposed position on the starboard side of the ship. One of his letters home described a Viet Cong rocket attack that occurred when he was in the process of opening a hatch. A shipmate had manually closed multiple "dogs" (fasteners) on the hatch to ensure maximum watertight integrity to prevent the spread of fire or flooding throughout the ship in the event of combat damage. Had the guy not so thoroughly secured it, Alan would have been passing through the opening, en route to the main deck from inside the ship, and been severely wounded or killed.

On another occasion, the *Iredell County* came under attack after leaving Dong Tam, a huge supply base, by Viet Cong armed with RPGs. Alan was manning a .50-caliber machine gun near the bridge house and was exposed to enemy fire. This letter made me very worried, and brought the war home to me in a very personal way. I did not worry as much about myself, during the war, as I did him. His ship came under fire so many times on inland waterways in Vietnam that he could not remember the details of them all.

In his letters home, Alan would periodically inquire about his '57 Chevy which he was letting me use. He knew that it had some serious wiring problems. (I occasionally had to have someone push it to help me get it started.) He also wondered if I liked the eight-track tapes that he had left in the car. Two of them were by Credence Clear Water Revival. I recall one of their songs in particular, "Fortunate Son." A part of the lyrics are: "It ain't me, it ain't me, I ain't no senator's son. It ain't me, it ain't me. I ain't no fortunate one, no." Many, if not most, of the men and women who went off to war were proud to serve. However, there was some resentment among draftees, who felt that they were not fortunate sons. Alan was released from active duty in April of 1970 and went to work for the family appliance repair business.

A MENTOR "CROSSES THE BAR"

In November of 2009 the world lost a great human being; Chico, a wonderful citizen; the Naval Reserve, a fine example of what a sailor should be; and I, a friend, when Chief Machinist's Mate John Hammons passed away. He was the Navy recruiter who had launched me into Navy life when on 13 September 1961 I became a member of the Naval Reserve Unit at the Chico Airport. After returning to Chico following my tour in Vietnam, I would run into John at every veteran's event I attended. He was always happy to see me and I him. Each time I vowed to ask him to coffee and reminisce about the old times, but, to my regret, I never followed up on those intentions.

Although retired, John always wore his chief's uniform at veterans' events and he was still able to fit into it. His sleeve had many, many hash marks on it, each one representing four years of service to the Navy. At Veterans Day ceremonies on 11 November 2009, I noticed that John was not there and asked Chief Yeoman Sharon Nichols, a fellow reservist, where John was. She commented that it was not like John to miss this event. Within a few days, the notice of his passing was in our local paper. I can picture John at the *Pearly Gates*, saluting and asking St. Peter, "Permission to come aboard, sir?" St. Peter, returning his salute, replies, "Permission granted. Well done, Chief!"

14

Hopewell Reunions

I wish there was a way to know you're in the good old days before you've actually left them.
—Andy Bernard

Now I remembered a captain's honor and his only duty: to bring his crew back alive.
—Carsten Jensen, *We, the Drowned*

Two decades after I left active duty, I spearheaded the first of many *Hopewell* reunions. Such affairs are largely as depicted in films. Former sailors—middle aged, wearing tropical shirts and typically ship's ball caps covering graying or thinning hair—reminiscing about the adventures of their youth. For those able to attend them, ship reunions provide opportunities to see old shipmates and, perhaps, to thank one's officers who collectively ensured the safety of one's ship. The most important individual being, of course, the commanding officer, who may have once seemed remote and a little forbidding, but as it turns out, is often a friendly, insightful individual.

I served under three different commanding officers while aboard the *Hopewell*. The first was Comdr. Robert Alvie Moore, who commanded the destroyer from May 1963 to September 1964. He was affectionately referred to as "Crazy Bob" by the crew. Reportedly, on more than one occasion, he made it a practice of eating peanut butter and banana sandwiches on the flying bridge while attired in pajamas. "Sugar Bob," another moniker assigned to him, also won prizes on San Diego area radio show contests—things like electric tooth brushes—and sold them to crew members. Of course, this was scuttlebutt relayed to me when I reported aboard and, as has been said more than once, "all sailors are liars."

My second captain was Comdr. Donald Fleming Milligan, aboard from September 1964 to September 1965. Though short in stature he had a commanding presence about him, and was very thorough in everything he did, even beyond that expected of all commanding officers. These attributes endeared him to me during my work for him as a yeoman. When he left the *Hopewell*, he became the commander of Escort Squadron Seven. He retired in the rank of captain at the culmination of a "stellar" naval career.

My third captain was Comdr. Leonard Howard Nettnin, who was in command from September 1965 to March 1968. He had a disarming smile, a commanding presence, and I had complete trust in him. It was not until a *Hopewell* reunion in Baton Rouge, Louisiana, in April 1995, that I learned many personal details about him from his widow Lee, including that he had been a passionate stamp collector, like me. Although we had talked of things other than Navy business while aboard the *Hopewell*, I never asked, nor did he tell me, of his interest in philately. This of course is common; we often do not know details about even very close friends until we read their obituaries, or attend their funerals. Sad, really.

For many years after I left active duty, I wondered, like many if not all former sailors, what had happened to my ship, and I continually checked the VFW magazine and other such veterans' publications to see if there was a reunion afoot. Not knowing the *Hopewell* had left naval service, I called the San Diego Naval Base to see if she was in port. She wasn't; the naval station dispatcher suggested that she might be laid up in "mothballs" in the "Ghost Fleet" at Bremerton, Washington.

While planning the *Hopewell* reunion in 1995, I found out my former destroyer was resting on the seafloor off San Clemente Island in southern California, sent to the bottom while serving as an unmanned target ship during a test of a new type missile. She did not accept her fate easily—after having survived three wars. Struck by a Walleye II missile, which detonated on impact, she remained afloat for two days. She was designated a hazard to navigation, because her 01 level stubbornly lingered above the surface of the ocean. A demolition team was dispatched on 11 February 1972 to put her down. It was foggy, wet and cold that day when, as the boat laden with explosives got to within 100 yards of the *Hopewell*, she slipped silently into the abyss. She sank on her own terms.

The first crew reunion took place in Baton Rouge, Louisiana. Part of the festivities took place aboard the destroyer *Kidd* (DD-661). This museum ship, a fully restored *Fletcher*-class destroyer was a sister ship to the *Hopewell*. To assist me in locating as many former *Hopewell* crewmembers as possible, I enlisted the help of two other men from different eras. Roger Burtness had served aboard the destroyer in

World War II and Bob Lawson during the Korean War. They were coordinators of their respective eras, and I was the Vietnam-era point person and the overall chairman.

Many of us wondered what had happened to the ship's bell after the *Hopewell* was struck from naval service. Certainly the Navy would have saved this revered item. A ship's bell announced to all aboard her and to the crews of nearby ships, the comings and goings of any and all commanders of importance. In the case of *Hopewell*'s commanding officer, the petty-officer-of-the-watch would ring the bell four times (denoting an officer of the rank of commander or captain) ding-ding, ding-ding and announce "*Hopewell*, arriving," followed by a single ding once he stepped aboard. This ritual signified that our commanding officer was aboard. Concurrently, the captain's absentee pennant flying from the mast would be hauled down. The same process with the words "Hopewell departing" told ship's company that the captain was leaving the ship. Was the bell in some dusty warehouse in Washington, D.C.? We learned that the bell was in a Naval Reserve Facility in San Jose. We requested and were granted permission to use it at the *Hopewell* reunion in San Diego in 2008.

HOPEWELL'S FINAL RESTING PLACE FOUND

In 2011, I received word from Ed Chew, a former supply officer on the *Hopewell* that her hulk had been found on the bottom off San Clemente Island. The following material is from an account by Kendall Raine regarding the discovery of the shipwreck:

> Weather conditions were perfect today as Capt. Ray Arntz, John Walker, Scott Brooks and I left Huntington Beach aboard *Sundiver II* bound for San Clemente Island. Our objective was to dive a promising target which Gary Fabian's extensive research of multi-beam sonar data and navy records indicated is the *Hopewell* (DD 681)....
>
> Launched May 2, 1943, *Hopewell*...saw service in the Pacific in WWII, Korea and Vietnam until decommissioned and struck on 2 January 1970. Her service in Korea earned her the nickname the "Duke of Wonsan" in reference to action in the port of Wonsan. Modified between the wars, she had her twin quintuple torpedo tubes and her number three aft mounted forward facing 5" main armament removed. Other modifications included a more robust mast to accommodate modern radar arrays. [She also had 3-inch/50 mounts installed, two amidships and one aft.]

Chapter 14

The sun was bright and seas flat due to the absence of wind at the dive site. As the boat carrying the men approached the dive site, it encountered sea lions floating on their backs with fins in the air. A few energetic ones played with the float on the end of the down line as the men "geared up." There was about a half knot of current that day and the estimated descent time was roughly five minutes. The target was believed to be lying on its side:

> John and I hit a thermocline [where the temperature of the water becomes progressively colder] at about 150' and the water became colder, clearer and darker. At roughly 30 feet above the wreck white and brown basket sponges came into view. We came in on the starboard side roughly amidships. As we leveled off at main deck level, John and I assumed our usual formation with him flying beyond me and inboard of the starboard rail with me just outboard. I was delighted to find the ship completely upright as this would make for a more interesting site than if she were on her side.... *Hopewell* had become a veritable garden with thousands of fish, metridium, sponges and Corynactis covering every inch of the hull.
>
> I looked inboard and upward to see the two forward main turrets with their barrels angled up. Against the cobalt blue of the shallower depths, this was majestic.... I rounded the bow and headed aft down the port side. No sooner had I made the turn than the ship's massive port anchor came into view.
>
> I continued aft as John made the run at the bow. I admired the impressive main guns and then looked up to see the angled bridge rising above the main deck. Just ahead of me was the port side hedgehog and above that was a row of intact portholes of the fighting bridge.
>
> I continued aft along the port rail and came across a large pipe bent outward by some massive force. Photos of the ship suggest this was railing on the second deck, now bent forward by the explosion of the Walleye missile. I maneuvered past this grating and found a massive hole in the main deck where the twin stacks once stood. This was obviously a missile impact area and the hole looked like it went at least two decks down....
>
> We noted the two aft main turrets appeared to have been removed and capped prior to the ships sinking. Together we motored aft through clouds of small fish only to be confronted by a steel wall where none should be. Rising at an angle perhaps 15 feet, we were seeing the fantail of the ship which had been thrust

upward by the massive power of the ships impact on the ocean floor. I rolled over the side and the port screw, completely covered in metridium and sponges came into view. With its three blades and pointed hub, it looked beautiful and threatening at the same time.

The transom looked strangely tapered and we rounded the stern. It was even more richly covered with sponges. We flew just over the starboard screw noticing the struts and rudders and then up again to the main deck. As our allotted bottom time ended John and I began our ascent....

Photo 14-1

Photograph of the barnacle-encrusted bow of the ex-*Hopewell* taken during the dive.

Postscript

Transferring you from the mess decks to ship's office turned out well for both of us. Many executive officers had real problems with their offices that I was spared. It was always satisfying to know that all the reports called for by the tickler file would appear on my desk for signature. You had a very good man and an excellent leading chief to teach you how ship's office should be run. I expect that since leaving the Hopewell *you have appreciated the skills you learned. Remember when we stopped in Agana, Guam on the way to Westpac and many of the Chief's relatives came to the ship?*

Reading the addresses of the Hopewell's *crew reminds me of how the military service acts to bring men from all over the nation into a very small space to work for a common patriotic goal. Living closely with guys from other regions does much to weld us together as a nation. The downsizing of the military is necessary, but I regret that a high proportion of men under forty have not had that experience.*

—Wayne Irwin, former *Hopewell* executive officer,
commenting in a letter, dated 4 July 1995,
on my service as a yeoman aboard ship.

I could not possibly recount in a single book all the crazy things that I encountered as a naive young man from a small hamlet in northern California. I entered the Navy as an immature young man with carousing on my mind. As I progressed through my enlistment and only gradually grew up, I became more drawn to experiencing tours of the cities/areas my ship visited, and less to devoting all my time to drinking and chasing girls. I feel that I matured in a way that I might not have, had I not been in the Navy.

As the years have passed, I have become increasingly proud of my Navy service, my time aboard the *Hopewell*, and of shared experiences with my shipmates. Conversely, I remember times of being very homesick and missing my family as one might expect, and of eagerly waiting to get out of the Navy and return to college. Most former sailors looking back on their time in the Navy, tend to remember the good times like the liberties and drinking with shipmates, visiting foreign ports, and exotic ladies—not the physically demanding and sometimes brutal duty aboard ship. For me, such challenges included long hours on the gunline off Vietnam, and carrying out my duties during fierce storms at sea, things that a normal person would not have to contend with in their daily lives.

As most Americans of a particular age are aware, the service of those of us who served in Vietnam was not usually lauded once we returned home. As has been widely reported, many veterans were spit upon and called "baby killer." While attending Chico State, I was a "weekend warrior," a member of the Naval Reserve, and wore my uniform on campus on Veterans Day and for other observances. This action caused some pretty intense looks from some of my fellow students. During my political science class, I related my experience in the Gulf of Tonkin and was ridiculed by the professor who disliked the American military and the war. Realizing that I could not change his perspective, I chose not to engage in further useless debate.

I would not trade my Navy experience for anything. I lived and worked closely with all my shipmates, went on runs ashore with many of them and enjoyed their company while exploring, cavorting and drinking. I loved the sea service and I was proud to wear the uniform. I was particularly proud of my ship and her crew. I applaud all those who served, those who are now serving, and those who will serve. I still tear up when hearing the National Anthem, and find it very hard to observe others who talk through it or otherwise disrespect our country. I must admit that I also feel great emotion when *Anchors Aweigh* or the *Navy Hymn* is played.

I also hold dear my post-Navy experiences. By leaving the service, I was able to return to college and study art under Dr. Janet E. Turner, a printmaker of world renown. I was also able to travel to Holland to spend time with M. C. Escher—in my estimation, the preeminent printmaker of the 20th Century—and have been making fine art prints ever since. For those readers particularly interested in fine art, I have provided (in the appendix) a summary of this experience, and also a few examples of my prints, which some people have (generously) compared to the work of Escher. Most importantly, had I stayed in the Navy, I would not have met and married Merry, my soul-mate.

I hope that this book helps readers who served in the Navy, Coast Guard, or the Merchant Marine, to recall with fondness, deep waters, runs ashore, and good shipmates. Vice Adm. Harold Koenig, USN (Ret.), captured the essence of being a sailor in the following poem:

ODE TO THE NAVY
I Like the Navy; Reflections of a Blackshoe

I liked standing on the bridge wing at sunrise with salt spray in my face and clean ocean winds whipping in from the four quarters of the globe—the destroyer beneath me feeling like a living thing as her engines drove her swiftly through the sea.

I liked the sounds of the Navy—the piercing trill of the boatswain's pipe, the syncopated clangor of the ship's bell on the quarterdeck, the harsh squawk of the 1MC, and the strong language and laughter of sailors at work.

I liked Navy vessels—nervous darting destroyers, plodding fleet auxiliaries and amphibs, sleek submarines and steady solid aircraft carriers.

I liked the proud names of Navy ships: *Midway, Lexington, Saratoga, Coral Sea, Antietam, Valley Forge*—memorials of great battles won and tribulations overcome.

I liked the lean angular names of Navy "tin cans" and escorts—*Barney, Dahlgren, Mullinix, McCloy, Damato, Leftwich, Mills*—mementos of heroes who went before us. And the others—*San Jose, San Diego, Los Angeles, St. Paul* and *Chicago*—named for our cities.

I liked the tempo of a Navy band blaring through the topside speakers as we pulled away from the oiler after refueling at sea.

I liked liberty call and the spicy scent of a foreign port.

I even liked the never ending paperwork and all hands working parties as my ship filled herself with the multitude of supplies, both mundane and to cut ties to the land and carry out her mission anywhere on the globe where there was water to float her.

Postscript

I liked sailors, officers and enlisted men from all parts of the land, farms of the Midwest, small towns of New England, from the cities, the mountains and the prairies, from all walks of life. I trusted and depended on them as they trusted and depended on me—for professional competence, for comradeship, for strength and courage. In a word, they were shipmates then and forever.

I liked the surge of adventure in my heart, when the word was passed: "Now set the special sea and anchor detail—all hands to quarters for leaving port," and I liked the infectious thrill of sighting home again, with the waving hands of welcome from family and friends waiting pier side.

The work was hard and dangerous; the going rough at times; the parting from loved ones painful, but the companionship of robust Navy laughter, the "all for one and one for all" philosophy of the sea was ever present.

I liked the serenity of the sea after a day of hard ship's work, as flying fish flitted across the wave tops and sunset gave way to night.

I liked the feel of the Navy in darkness—the masthead and range lights, the red and green navigation lights and stern light, the pulsating phosphorescence of radar repeaters-they cut through the dusk and joined with the mirror of stars overhead. And I liked drifting off to sleep lulled by the myriad noises large and small that told me that my ship was alive and well, and that my shipmates on watch would keep me safe.

I liked quiet mid-watches with the aroma of strong coffee—the lifeblood of the Navy permeating everywhere.

And I liked hectic watches when the exacting minuet of haze-gray shapes racing at flank speed kept all hands on a razor edge of alertness.

I liked the sudden electricity of "General quarters, general quarters, all hands man your battle stations," followed by the hurried clamor of running feet on ladders and the resounding thump of watertight doors as the ship transformed herself in a few brief seconds from a peaceful workplace to a weapon of war—ready for anything.

And I liked the sight of space-age equipment manned by youngsters clad in dungarees and sound-powered phones that their grandfathers would still recognize.

I liked the traditions of the Navy and the men and women who made them. I liked the proud names of Navy heroes: Halsey, Nimitz, Perry, Farragut, John Paul Jones and Burke. A sailor could find much in the Navy: comrades-in-arms, pride in self and country, mastery of the seaman's trade. An adolescent could find adulthood.

In years to come, when sailors are home from the sea, they will still remember with fondness and respect the ocean in all its moods—the impossible shimmering mirror calm and the storm-tossed green water surging over the bow. And then there will come again a faint whiff of stack gas, a faint echo of engine and rudder orders, a vision of the bright bunting of signal flags snapping at the yardarm, a refrain of hearty laughter in the wardroom and chief's quarters and mess decks.

Gone ashore for good they will grow wistful about their Navy days, when the seas belonged to them and a new port of call was ever over the horizon.

Remembering this, they will stand taller and say "I WAS A SAILOR ONCE. I WAS PART OF THE NAVY & THE NAVY WILL ALWAYS BE PART OF ME."

Appendix: My Time with Escher

By chance, I saw an image of a print in one of the Bay Area newspapers in 1969 and that print was *Day and Night* by M. C. Escher. I knew instantly that this was my "Holy Grail" and Escher was my Sage. I was a graduate art student, with a printmaking emphasis, at California State University, Chico. Later I met and got to know David Trufant, another graduate printmaker. He and I made a trip to the Vorpal Gallery in San Francisco to see an Escher exhibit, at which I could have purchased Escher's *Day and Night* print for $300. I wanted that print, but only had $400 to fund planned travel to Europe to hopefully meet Escher in person. (Recently, that print sold for $56,000.)

David, his wife Ellen, and I travelled to Europe in June 1970. After our plane landed at Schiphol International Airport in Amsterdam, I went to a phone booth, looked up M. C. Escher's name in the phone book, and called him. I could hardly believe that his number was listed, as at that time artists the world over wanted to speak with or visit him. I said "Mr. Escher my name is Michael Halldorson. I am a graduate printmaking student from Chico, California and I must see you."

He replied "I just got rid of two Americans," but then consented to give me fifteen minutes of his time, at three o'clock that coming Friday afternoon. I was ecstatic. That Friday, I took the Eurorail train from Amsterdam to the town of Baarn, where Escher lived. I arrived in Baarn hours early, sat in a quaint café with a cup of cappuccino and fantasized what it would be like meeting the great print maker. Could I possibly fit all my questions for him into that amount of time, or would we just get started before my time was up? A taxi delivered me to Escher's home fifteen minutes early. At precisely three o'clock I knocked on his door.

I was greeted at the door by a woman who I assumed was Escher's wife, but learned later was the housekeeper with a friend over for tea. The two women returned to their visiting and a slight, white-haired man then came into the room and took me back to his studio. I was carrying the best print I had done as an artist, a framed copy of *Pollution I*. Escher looked at it and complimented me; although today, and even then, I thought he must have been very generous in his assessment of my work. I had taken the print to Europe with the intent of presenting it to Escher. How presumptuous of me to do so with no way of knowing if I would even get a chance to see him.

Photo Appendix-1

Michael Halldorson's print *Pollution I*, lying on Escher's work table.

He asked if I would be interested in seeing his latest print and the wood blocks that he carved for it. My response was an emphatic "yes!" The print was *Snakes*, which would be the last art he ever produced. He unwrapped the carved blocks from their protective wraps and revealed to me how they were used in the print making process. I was enthralled—amazed at his work—and commented that the print was perfect. He then pointed out one line that did not match up perfectly with its counterpart on the block. Each block was a third of a circle. They were printed by rotating each inked block 120 degrees to connect with the previously printed one. In the middle of the print was a pin hole which he used to rotate the blocks. Each block produced one color and each block was printed three times, so there were nine separate printings to produce his print. It would have taken a magnifying glass for most people to notice the "imperfection" Escher pointed out. To him, it was a glaring error.

I asked Escher if he would accept the gift of my print. He kindly declined my offer with the explanation that he was moving to what I believe was some kind of artists' residence. He told me that he could not take all of his possessions, let alone my print. During my visit, I took four photographs of his studio. Unbeknownst to me, my print was lying on his work-surface in the foreground as I took these pictures. Today I am inordinately proud that for at least a very short time, my work shared the same space as his. I regret not taking more photos of the studio that day, or even around the neighborhood, and also that I did not ask the housekeeper to take my picture with him. The only explanation that gives me any degree of solace is that I was in the presence of greatness and was overwhelmed. Memories diminish over time but I seem to recall that his home was red brick on a shaded tree-lined street.

The fifteen minutes allotted to me turned into several hours as we discussed techniques, materials, philosophies, etc. When Escher told me that he was sorry that he could not accept my print, he also said that he was regretful that he could not in turn give me one of his prints as they were not his to give. He used the expression "treasures-of-the-state" or words to that effect. I replied "Oh, that's okay." I commented, "Mr. Escher you are really big in the United States." He said "Yes I am, aren't I." There was no conceit in his answer—he was merely stating a fact.

At one point in our conversation, I offered to stay in Holland and print his prints for him, if he would just pay me a minimum wage. He seemed a little amused, and told me that he alone printed his work. He was referring to his woodcuts, not his lithographs. Professional lithographers printed these prints as he stood over the press personally monitoring the quality. He would destroy any "pull" that did not meet his very high standards.

I am forever grateful to Escher for his allowing me a portion of his precious time. I was so in awe of him during the meeting that I did not think to have him autograph my sketchbook. During my subsequent stay in Amsterdam, I went into a gallery and could have purchased his *Day and Night* for $127, but the slim funds I had needed to last until my return to the United States. My current hope is that one day I will be able to afford an Escher print.

Maurits Cornelius Escher passed away on March 27, 1972 less than two years after my visit. I like to believe that he has opportunity in heaven to create all the prints that he could not get on paper during his time on Earth. People have commented over the years that some of my prints look like Escher's. Where I am deeply flattered, mine don't even approach his work; they merely reflect his influence. Escher once observed:

> If only you knew the things I have seen in the darkness of night, at times I have nearly been driven mad at being unable to express these things in visual terms. In comparison with my visions, every single print is a failure and reflects not even a fraction of what might have been.

I will always treasure greatly my time with M. C. Escher.

Photographs of two of my prints are displayed on the following two pages:

Photo Appendix-2

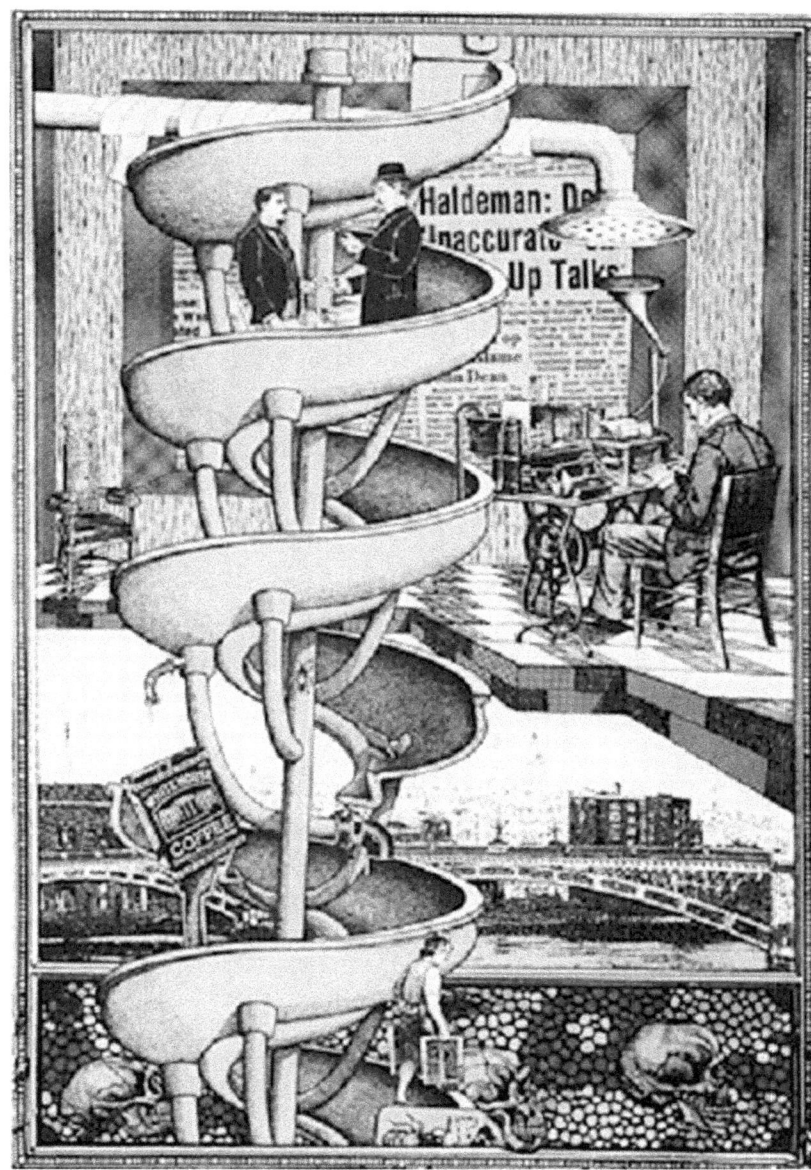

Print *At That Point in Time* – serigraph – produced in 1973

Appendix

Photo Appendix-3

Print *Babylonian Pipe Dream* – serigraph – produced in 1974